NORTH CAROLINA
STATE BOARD OF COMMUNITY COLLEGES
LIBRARIES
ASHEVILLE-BUNCOMBE TECHNICAL COMMUNITY COLLEGE

DISCARDED

JUN 25 2025

Mechanical Hands
Illustrated

Mechanical Hands Illustrated

Revised Edition

Edited by

Ichiro Kato

*Department of Science and Engineering
Waseda University, Japan*

Kuni Sadamoto

*Survey Japan
Tokyo, Japan*

● **HEMISPHERE PUBLISHING CORPORATION**
A subsidiary of Harper & Row, Publishers, Inc.

Washington   New York   London

DISTRIBUTION OUTSIDE NORTH AMERICA

SPRINGER-VERLAG

Berlin   Heidelberg   Paris   New York   Tokyo   London

MECHANICAL HANDS ILLUSTRATED: Revised Edition

Copyright © 1987 by Hemisphere Publishing Corporation. All rights reserved. Printed in the United States of America. Except as permitted under the United States Copyright Act of 1976, no part of this publication may be reproduced or distributed in any form or by any means, or stored in a data base or retrieval system, without the prior written permission of the publisher.

1 2 3 4 5 6 7 8 9 0 EBEB 8 9 8 7 6

The original Japanese version, compiled by Ichiro Kato, was first published by Kogyo Chosakai Publishing Co., Ltd.

This book was typeset by Hemisphere Publishing Corporation. The editors were Mary Prescott and Elizabeth Maggiora; the production supervisor was Miriam Gonzalez; and the typesetter was Peggy M. Rote.
Edwards Brothers, Inc. was printer and binder.

Library of Congress Cataloging in Publication Data
Katō, Ichirō, date.
   Mechanical hands illustrated.

   Translation of: Zukai mekanikaru hando.
   Includes bibliography and index.
   1. Manipulators (Mechanism) 2. Robotics. I. Title.
TJ211.K36613 1985   629.8'92   84-27906
ISBN 0-89116-374-3   Hemisphere Publishing Corporation

DISTRIBUTION OUTSIDE NORTH AMERICA:
ISBN 3-540-15429-9   Springer-Verlag   Berlin

Contents

Preface	xiii
Preface to English-Language Edition	xv
Preface to the First Edition	xvii

PART 1: INTRODUCTION

CHAPTER 1   Current Trends in Research and Development on Artificial Hands   3
Ichiro Kato

Functioning of Hands	3
Robot Hands	4
Prosthetic Arms	4
New Applications	5
References	6

CHAPTER 2   Mechanism of Artificial Hands   7
Kazuo Tanie

Classification of Artificial Hands	7
Degrees of Freedom of Artificial Hands	8
Mechanism	9
Sensors	13

CHAPTER 3   Analysis of the Operation of Mechanical Hands   14
Hideyo Ito

Role of Hands in Automation of Manual Work	15
Classification of Tasks in Relation to Fixing	15
Analysis of the Hand's Operation	16

	Problems in Designing Modular Hands	21
	References	21
CHAPTER 4	Grasping in Handwork *Hideyo Ito*	22
	Restraint of the Object of Handwork	22
	Hand Restraint and Grasp	23
	Meaning of Grasping	24
	Function of Grasping	24
	Analysis of the Grasping Function of Human Hands	25
CHAPTER 5	Characteristics and Design of Power Sources *Toyokazu Mitsuoka*	29
	Characteristics of Representative Power Sources	30
	Power Sources for Self-Moving Robots	33

PART 2: DRAWINGS

HUMAN TYPE

Chuo Hand X1	37
Flexible Mechanical Hand	38
WAM-4	39
WAM-5	40
Waseda Hand 5	41
All-electric Prosthetic Hand (TD50-3)	42
TD-2 Hand	43
All-electric Mechanical Hand (TDH51-1)	44
Electric Prosthetic Forearm (TDU51-1,2)	45
Human-type Hand	46
Waseda Hand 4	47
Waseda Imasen Myoelectric Hand	48
Kumadai Hand IV	49
New 10H7 with 7 Degrees of Freedom Powered by Oil Pressure	50
Prosthetic Hand with 7 Degrees of Freedom at the Shoulder Powered by Oil Pressure	51
Prototype of Prosthetic Upper Arm Powered by Oil Pressure	52
Prosthetic Forearm with 3 Degrees of Freedom Powered by Both Electricity and Oil Pressure	53
Prosthetic Hand with 7 Degrees of Freedom Powered by Oil Pressure (Prototype 1)	54
Instrument for Finger Powered by Gas Pressure	55

FREE TYPE: PICKING

Hand with Three Fingers	56
External Grasping Swing Lever Type Hand	57
TDG-1 Hand	58
Hand for Upsetting	59
Removable Hand	60
Double Hand (for Machine Tools)	61

CONTENTS vii

Hand for Forging Work	62
Gripper for Two Articles at a Time	63
Gripper for Gear Conveying	64
Hand for Drill Loading	65
Pick and Place Unit 1	66
Pick and Place Unit 2	67
Slave Hand 1RL	68
Slave Hand 2R	69
Hand for Loading and Unloading	70
Oil Pressure Hand with Screwing	71
Oil Pressure Hand for Conical Steel Material	72
Hand for Funnels	73
Grip Hand for Vertical Picking and Horizontal Setting	74
May Hand MIII	75

FREE TYPE: SNAP

Chuo Hand X2	76
U.I. Hand	77
TDL Dual Hand	78
High-quality Manipulator	79
ETL Manipulator	80
External Grasping-type Hand with Floating Swing Levers	81
External Grasping and Holding Hand	82
External Grasping and Holding Hand	83
MEL Arm	84
Artificial Hand and Arm Controlled by Electricity and Oil Pressure	85
Multijointed Robot Hand (Type KAR-2)	86
Hand for Marine Work	87
MEL Hand	88
Oil Pressure Grip	89
Oil Pressure Grip	90
Oil Pressure Hand for Copper Wire Bobbins	91
Oil Pressure Hand for Wheel Axles	92
Clamp Hand	93
Hand for Columns	94
Hand for Weight Lifting Inversion	95
Oil Pressure Hand	96
Standard Hand for Kawasaki Unimates	97
Parallel Movable Finger	98
Hand for Rear Head Grinding	99
Three-direction Grip Hand	100
Parallel Grip Hand	101
Single Three Arms plus Jaw Twisting Unit (R-4)	102
Clamp Jaw	103
Clamp Jaw	104
Hand for Stacking Aluminum Ingots	105
Hand for Carrying Copper Pipes	106
KHR-800W	107
Hand for TV Frame	108
Hand for Brown Tubes	109
General Utility Hand	110
Mechanical Hand with Vacuum Chuck	111

FREE TYPE: GRIP

Hand with Chucks on Both Sides	112
Soft Gripper	113
Hand for Front Heads	114
Oil Pressure Hand for Cylindrical Steel Material	115
Gripper for Pulling Out in Die Casting	116
Downward Machine Fingers for Transferring	117
Machine Fingers for Transferring	118
High-speed Machine Fingers for Transferring	119
Embracer 3,4	120
Artificial Hand Using Windthrow Effect	121

FREE TYPE: SUCTION

Thin-type Vacuum Chuck	122
Suction Chuck for Thin Plates	123
Automatic Tracking Vacuum Chuck	124
Vacuum Hand (Ejector Type)	125
Vacuum Hand (Blower Type)	126
Vacuum Hand	127
Vacuum Suction-type Hand	128
Hang Arm I	129
Hang Arm II	130
Double Venturi-type Jaw	131
Double Venturi-type Jaw	132
Hand for Conveying Brown Tubes	133
Suction Hand with Jaw	134
Gripper for Bowling Balls	135
Vacuum Hand	136

FREE TYPE: MAGNETISM

Magnetic Hand	137
Magnetic Hand	138
Hand for Transferring Rotors	139
Modular-type Magnetic Hand (Type KMR-2)	140
Magnetic Jaw	141

FREE TYPE: OTHERS

Inner Grasping Swing Lever Hand	142
Wiping Hand	143
Hand for Working on Rear Head Oil Holes	144
Gripper for Press Loading	145
Swing Hook	146
Hand for Automatic Arc Welding	147
Welding Gun Autochanger	148
Hand for Arc Welding	149

PART 3: PATENTS

FREE TYPE: PICKING

Plural Pick-up Hand	153
Hand with Independent Pawls	153

Three-pawl Hand	154
Fingers with Work Sensor	154
Grasp Hand for Multiple Works on Conveyor	155
Grasp Fingers for Large-scale Bodies	155
Fingers with the Ability to Recognize Work	156
Hand with a Couple of Clamp Arms	156
Hand with Three Concentric Fingers	157
Grasping Manipulator for Plural Goods	157
Hand Powered in Common with the Transfer Arm	158
Snapping Device	158
Hand for Sliding Cylinders in Two Positions	159
Grasping Device for Materials	159
Hand with a Leaf Spring Switch Mechanism	160
Transfer Pawls for Nuts	160
Transfer Device in the Right and Left, Up and Down Directions	161
Snap Device for Objects such as Spoons and Dishes	161
Snap Instrument for Cube Sugar	162
Work Gripper	162
Hand with Snap in the Guide Frame	163
Hand with Removable Grasp Frames	163
Grasping Hand Movable Up and Down and Rotationally	164
Transport Hand for Work on Turning Machine Tool	164

FREE TYPE: SNAP

Safe and Reliable Hand	165
Grasping Device with Probe Directly Touchable to Work	165
Probe Finger into Grasp Pressure and Displacement	166
Rapidly Changeable Chuck	166
Hand for Objects of Complex Shape	167
Hand with Snap Sensors	167
Multijoint Gripper	168
Hand Adaptable to the Weight, Shape, and Stiffness of Materials	168
Hand with Changeable Grasping Force and Speed	169
Parallel Links Switched by One Cylinder	169
Grasping Hand for Narrow Space	170
Fingers Confirming the Grasp of Thin Plates	170
Fingers with Reversible Long and Short Levers	171
Finger with Conductive Probes	171
Fingers Adjustable to the Displacement of Objects	172
Grasping Hand for Cylindrical Fiber Work	172
Finger with Resistance Powder Sensor	173
Hand with Two Types of Finger	173
Hand for Setting the Center of Objects	174
Finger for Grasping the Center of Objects by Contact	174
Center Holding Hand with Sensors	175
Grasping Hand for Narrow Space	175
Finger for Modifying the Declination of Objects	176
Three Fingers with an Air Nozzle for Cleaning	176
Fingers for Composition with Sensor for Fitting Direction	177
Slide Hand	177
Grasping Hand for Large-scale Materials	178
Hand for Setting the Center of Objects	178

Parallel Movable Finger	179
Hand with Adjustable Pressure	179
Robot Hand for Compact Working Area	180
Hand with a Cam and a Spring	180
Fingers for Cylindrical Objects	181
Grasping Device with a Couple of Horizontally Movable Jaws	181
Easily Removable Hand for Various Cylindrical Objects	182
Grasping Device	182
Work Grasping Device	183
Fingers with Different Concave Parts	183
Hand with Small-diameter Grasping Instrument	184
Hand for Shaft-type Objects	184
Grasping Device for Manipulators	185
Grasping Device for Various Objects	185
Hand with Rack and Pinion	186
Rotational Grasp Hand	186
Hand with Normal Screw and Reversing Screw	187
Grasping Hand with Magnets	187
Remote-control Grasping Instrument	188
Hand Powered by the Weight of Objects	188
Hand for Exchange of Tools	189
Hand with Chucks for Cylindrical Objects	189
Hand with Electric Revolution Mechanism	190
Hand with a Fixed Pawl and a Movable Pawl	190
Electric Grasp Hand	191
Hand with a Free Rotational Spring Pawl	191
Hand with a Free Rotational Pawl	192
Grasping Hand for both Inner Diameter and Outer Diameter	192
Hand with Two Cylinders	193
Hand with a Rectilinear Pawl and a Rotational Pawl	193

FREE TYPE: GRIP

Fingers for Modifying the Disagreement of Centers of Different Rods	194
Fingers of Flexible Material	194
Hand with a Self-lock Mechanism	195
Push Touch Hand with an Oblique Guide	195
Fingers for Objects of Various Scales	196
Hand for Inserting Pipes	196
Centering Fingers of Guide Body	197
Fingers Holding the Center of Rods with Different Diameters	197
Fingers with a Relief Mechanism	198
Hand Switched by Pressure of Touching the Ground	198
Hand for Easy and Reliable Grasping of Large-scale Materials	199
Clamp and Pick-up Device	199

FREE TYPE: SUCTION

Suction Hand Movable Forward and Backward	200
Suction Hand with Ready Knockout Pin	200
Suction and Pull Hand	201
Suction Hand with Spool Containing Pressure Room	201

FREE TYPE: MAGNETISM

    Pulling Finger with Magnets — 202
    Magnetic Hand with Multiple Absorption Disks — 202

OTHER TYPE

    Hand Stretching from Inner Side — 203
    Fingers with Pressure Welding Materials — 203
    Hand Indifferent to Slight Attitude Change — 204
    Spot Welding Hand with Direction Control — 204
    Hand for Taking Out Cylindrical Materials — 205
    Universal Hand for Doughnut-like Objects — 205
    Grasping Hand for the Inner Stage of Work — 206
    Hand with Adjustable Gripping Center — 206
    Pullout Hand for House Boxes — 207
    Grasping Hand Consisting of Multiple Joints — 207
    Hand with Rotation-tunable Couple of Fingers — 208
    Hand with a Machine Mechanism in the Main — 208

Appendix: Actual Mechanical Hands — 209
Index: Manufacturers and Power Sources — 215

Preface

In the field of robotics, considerable effort is devoted to the development of effective multiple prehension manipulator systems. Most of these efforts have been directed toward attempts to duplicate certain functions of the human hand for industrial and medical applications.

The human hand is an intricate and complex system capable of a multitude of sensory and actuation functions. It has approximately 20 degrees of freedom controlled by a large number of muscles compared to only 7 degrees of freedom for the arm and wrist. Since the primary function of a mechanical hand is to grasp or pinch different objects, not all the degrees of freedom are necessary for every robot or prosthetic device.

Most practical tasks for manipulators have their special requirements and constraints which influence the design of their end effector. Consequently various hand configurations have to be designed to effectively provide the required functions for a particular application without additional complexity. The control of these functions is accomplished by suitable actuators. Miniature touch and pressure sensors which can be as small as 3 mm in size are used to control the grip.

This book is a valuable addition to the knowledge base in the field of robotics. It provides an excellent compilation of photographs, scaled drawings, and source material for a wide variety of mechanical hands for the user to choose from. Professor Ichiro Kato and Mr. Kuni Sadamoto have made a significant contribution to the technology transfer in this important field by collecting, translating, and editing the material in this book.

It is hoped that this American Edition matches the huge success of the version published in Japan and serves the needs of the rapidly growing English speaking community of manufacturers, designers, and users of robots and manipulators.

Department of Mechanical Engineering   Ali Seirig
University of Wisconsin-Madison

Preface to English-Language Edition

When this book was first published in Japan in 1977, it was regarded with high esteem. It is a great pleasure to me, as editor, that this book has been reprinted again and again because of readers' demands.

In fact, we have even received a great many orders for the Japanese version from readers overseas. This is probably because the book places emphasis on illustrating hand mechanisms, and so could appeal to foreign readers despite the language barrier. I believe, however, that the greatest reason for its popularity is that there is no other book of this kind in Japan or elsewhere despite the great potential need for it.

We expect that the 1990s will be the era of robots. I hope that this book will be widely used by the many robot designers and users throughout the world.

Ichiro Kato

Preface to the First Edition

Among all living things, only man has hands. Man's technology and culture came about because humans learned to use their four limbs separately, and to use their hands and feet for different purposes. It is said that the hand is the reason why man's intelligence became greater than that of any other animals; thus the hand is closely related to the brain.

Machines with hands and brains that imitate those of humans have come to be called robots. The industrial robot Verthtran was demonstrated for the first time 10 years ago at Harumi wharf in Japan. Since then, there has been an explosive development of robots in this country which has attracted the attention of the world because it seems to presage the future of robots.

At the present state of development, it is clear that both the hands and brains of robots must be improved so that they can develop into more advanced machines. It is evident that intelligence cannot be separated from the use of the hands, and the hand is one of the most important elements in the design of robots and in their selection by users.

In this book we have collected typical robot hands from Japan and other countries and illustrated their design and selling points through some 255 drawings and photographs from their manufacturers. To assist readers in using these illustrations, the recent developments in robots, the operation of robot hands, the working analysis of hands, and so on are explained by experts in these areas. In addition, the illustrations include drawings from domestic patents, and at the end of the book there is an index of the illustrations by manufacturer, robot type, and power source type.

Our own hands have infinite possibilities. The robot is an engineering approach to human hands. The aim of this book will be fulfilled if designers and users can obtain from it ideas for imitating the infinite possibilities of hands.

Ichiro Kato

Part 1

INTRODUCTION

Chapter 1

Current Trends in Research and Development on Artificial Hands

Research and development on artificial hands has been mainly along two lines, corresponding to their application as robots in industry, and as prosthetic devices for upper-limb amputees. The former application has led to a function-oriented design related to the tasks performed, while the latter has demanded an appearance similar to that of human hands as well as functions that can be substituted for those of upper limbs.

Research on artificial hands has been carried out in Japan for the past 10 years, beginning with the activities of a research group which led to the present Biomechanism Society. This book reviews the developments in artificial hands that have taken place over the past decade, summarizes the present state of the art, and makes some projections of likely future developments in this field.

The word "hand" is used in this book in two senses: to refer to the whole upper limb, and to refer to the part of the upper limb from the wrist to the tips of the fingers. An upper limb may be described as consisting of an arm, a wrist, and a hand. From the standpoint of functions, the arm performs the positioning of the hand, while the wrist joins the arm and the hand and determines the direction of the hand. The hand itself has the greatest influence on function, whether in a robot or a prosthetic arm. This can be seen clearly if we consider that in a mechanical model of a human upper limb, with some 27 degrees of freedom, 20 degrees of freedom are attributed to the hand. The design of the arm is reduced to the selection of type and location of moments, which have 6 or 7 degrees of freedom and are determined by the requirements for power and control. This design is not as versatile as that of the hand itself, on which we focus in the following chapters.

FUNCTIONING OF HANDS

Human hands consist of fingers and a palm. The great range of functions that can be performed by human hands is due to the flexibility given to the fingers by their joints, the many possible combinations of the fingers with each other and with the palm, and the cutaneous sense organs, which respond to stimuli such as pressure. It might be said that the unique combination of fingers, palm, and sense organs has great functional flexibility due to the great mechanical flexibility of the hand.

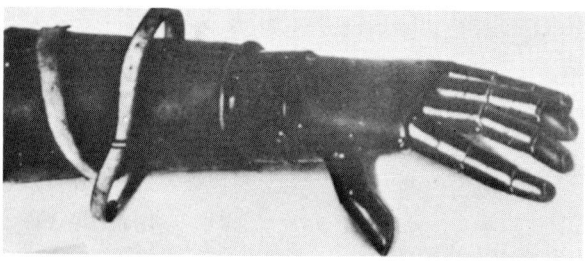

Figure 1 Prosthetic hand by Gotz.

Artificial hands are used as industrial robots in two ways: as an instrument or tool, and as a hand for grasping a tool. Up to now they have most often been used as tools, and it is common to substitute single-function robots for human hands to perform particular work functions.

In the case of prosthetic arms, the hands completely take the place of human hands. While the intact human may think mainly of the functioning of the hands, amputees are also greatly concerned with their appearance. Therefore, as illustrated by the prosthetic arm made of steel at Gotz von Berlichigen (Fig. 1), the design of prosthetic arms puts emphasis on appearance as well as function.

ROBOT HANDS

The basic function of hands is to grasp things with the fingers. Therefore, almost all industrial robot hands consist of a pair of tongs corresponding to nonjointed fingers and used for grasping. This type of robot exists in a variety of shapes, mechanisms, and capabilities. For instance, in one case a vacuum pad is used as a palm. These single-function hands have shapes that depend on the purpose for which they are used; hence many hands must be prepared in advance for different purposes. In some trials joints have been put in the fingers of industrial robots, but they do not necessarily utilize the function of joints.

In a recent trend, hands have been designed not as a tool, but as a hand supporting a tool. In this case, the hand may be designed so that the tips can easily be replaced by various tools used as attachments. The hand itself is designed to grasp the tools accurately. This type of robot has appeared in both Japan and the United States. Figure 2 shows an experimental robot of this type prepared for the NASA program. The difference between this type of hand and the hand used as a tool lies in whether replacements can be made at the wrist or at the fingers. In the latter case, tools grasped by the fingers generally do not have movable parts.

PROSTHETIC ARMS

When we consider human hands as mechanisms, we are impressed by their beauty of creation. Imitating them with machines and making artificial hands that are similar to real human hands is a great challenge. The design of prosthetic arms is determined by four conditions: function, size, weight,

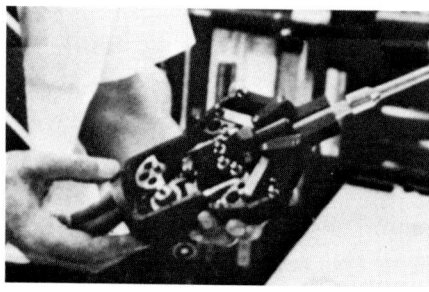

Figure 2 Hand tip for robots designed for NASA. Various tools are easily attached to the hand.

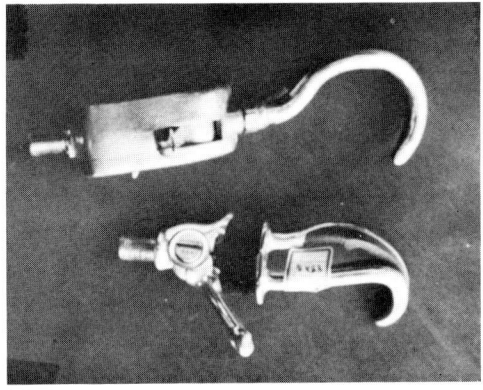

Figure 3 Hooks may be used to hang things from, to hold a pencil, and so on.

and appearance. The process of prosthetic arm development can be said to be similar to mountain climbing, where one approaches higher summits step by step.

Hooks, which have been widely used for a long time, are a kind of tongs (see Fig. 3). The current trend, however, is to make artificial hands that are similar to real human hands. Many commercial prosthetic arms have 1 degree of freedom and are capable of pinching with the thumb, index finger, and middle finger, or even with the five fingers for appearance. They can move only at the joints between the fingers and the palm and have no joints in the fingers.

The new trend in prosthetic arm development is toward multiple functions, providing each finger with joints, and providing freedom of rotation as well as extension. The goal is to approach the adaptability of human fingers. Using the supermicro realization technique, experimental trials have been performed with hands in which each finger is actuated independently by a tiny motor located in the palm (see Fig. 4).

NEW APPLICATIONS

Industrial robots have been widely used not only in the industrial sector, but also in space and ocean exploration. A new field of application is that of medical systems. In this case, the applications can be classified in three categories: diagnosis, operation, and nursing aid.

In diagnostic procedures relying on touch, the doctor's hand may be replaced by an artificial hand—for example, in diagnosing breast cancer or taking x-ray pictures under pressure. Operations at distant sites may be performed by artificial hands remotely controlled by doctors in a city. Finally, artificial hands may be used to perform laborious work required in the nursing care of humans—for example, in aiding heavily handicapped persons. The big difference between such medical applications and more conventional applications of artificial hands is that in the medical uses the objects are not hardware but human patients. Therefore, for such uses the development of hands with artificial cutaneous sensations is a strong requirement so that patients will not be injured.

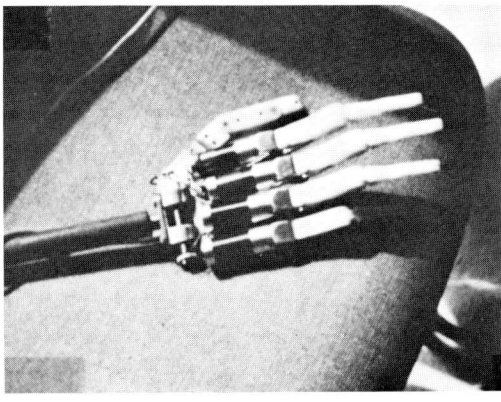

Figure 4 Telemechanic hand. Supersmall motors are placed at the base of each of the five fingers.

The development of artificial cutaneous sensations for future artificial hands is indispensable. Currently, the hardware developed for this purpose consists of arrays of rectangular sensors 3 mm long, located in 25 rows and 25 columns. The software being developed to give the sense of hardness to artificial hands has started by using human cutaneous sensations. These sensory devices, as well as being necessary for medical applications, will also lead to improved artificial hands for more conventional uses.

REFERENCES

1 I. Kato (translator), *Control of Human Limbs,* p. 14, Gakuken-Sha, 1968.
2 I. Kato (ed.), *Control of Human Limbs,* 3d ed., p. 97, Gakuken-Sha, 1977.
3 I. Kato (ed.), *Control of Human Limbs,* 3d ed., p. 37, Gakuken-Sha, 1977.
4 Special Issue on the Measurement of Hardness of the Living Body and Artificial Sensories, *Instrumentation and Control,* vol. 14, no. 3.

Chapter 2

Mechanism of Artificial Hands

Artificial hands have been classified as prosthetic arms or robots, depending on the main purpose for which they are used. Prosthetics are used as substitutes for missing human limbs, robots for automating manufacturing processes and freeing humans from labor in dangerous environments. The different demands associated with these uses give rise to different problems and considerations in the development process.

Since robots are used as a substitute for humans working with machines, the optimal design for their structure and appearance can be based on an accurate analysis of the functions being replaced. Prosthetic arms must be acceptable to amputees physically, physiologically, and psychologically, so that in addition to improving their functions, light weight, improved maneuverability, and geometric similarity to human hands are considered important in their design. For these reasons prosthetic arms and robot hands are different from each other structurally and geometrically and have their own unique features. In this chapter the mechanism of artificial hands is described.

CLASSIFICATION OF ARTIFICIAL HANDS

Prosthetic arms and robots may be classified as shown in Fig. 1. Cosmetic arms are made only for appearance and do not have any other functions. Utility arms have hooks at the tip for some work,

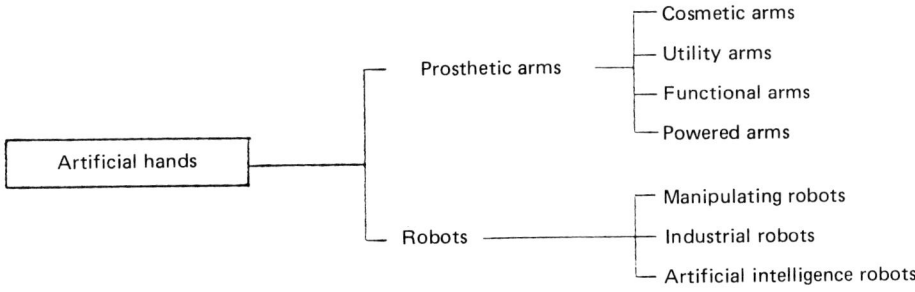

Figure 1 Classification of artificial hands.

but have no movable parts. Functional arms have terminal devices such as elbows or hooks which are controlled through cables from a harness by making use of the capability left to amputees. Powered arms have an external power device which can be controlled by some biophysical signal such as a myoelectric potential; they are intended to increase the functions of conventional prosthetic arms. Depending on the condition of the amputated parts, prosthetic arms may be referred to as forearm, upper arm, or whole arm prostheses.

There are three types of robot. Industrial robots are automatic manipulators developed for the purpose of labor reduction and automation in production lines; depending on the control method used, they are referred to as continuous path (CP) type, point to point (PTP) type, or a simplified type. A manipulating robot is a type of robot directly controlled by humans, such as the master-slave type used in nuclear reactors or the manipulators used on deep-sea submarines in ocean research. Artificial intelligence robots have been developed as experimental tools for analyzing biophysical functions and intelligence from the viewpoint of biomechanism; they are generally manipulators which can move autonomously under the control of large computers.

DEGREES OF FREEDOM OF ARTIFICIAL HANDS

Regardless of the purpose for which artificial hands are used, their design is modeled on the human hand. Therefore, before describing the mechanism of artificial hands, we should consider the structural model of human hands.

According to A. Morecki, a human upper limb can be modeled on the basis of anatomical structures as shown in Fig. 2. Here the Arabic numerals represent bone structures and the Roman numerals represent joints. If we denote the number of movable parts by n and the number of moments with f degrees of freedom by Pf, then the total number of degrees of freedom of the three-dimensional linked mechanism is given by

$$F = 6n - \sum_{f=1}^{5} fPf$$

where $f = 1, 2, \ldots, 5$.

For the model in Fig. 2, $n = 19, P_2 = 1, P_3 = 2, P_4 = 6$, and $P_5 = 11$, giving $F = 27$. For this number of degrees of freedom a total of about 50 muscles is used in connection with the joints. The total number of degrees of freedom for a hand, defined as shown in Fig. 2, is 22 for $n = 16, P_4 = 6$, and $P_5 = 10$.

The computational result above indicates that producing an accurate artificial arm requires 27 actuators, each of which has 1 degree of freedom. Furthermore, the hand has the greatest

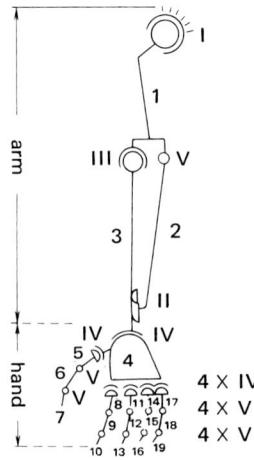

Figure 2 Structural model of the human upper limb.

concentration of degrees of freedom and requires 22 actuators. At present, it is nearly impossible in terms of dimensions, weight, and control to build artificial hands with this many degrees of freedom. In particular, from the standpoint of geometric shape and maneuverability, it is entirely impossible to build prosthetic arms. (Some exceptions to this statement exist, notably hands made by P. Rabishong.) Usually, by focusing on the functions, the degrees of freedom concentrated in the hands can be simplified.

For the function of grasping, 12 different geometries can be defined. These may be classified into two types according to the relation of the thumb and the other four fingers. In the first type, the two-point pinch, the thumb moves against one of the other four fingers and can do so from either side of them. The second type, the three-point pinch, involves the tips of the thumb, the index finger, and the middle finger, and is considered of secondary importance based on observations of ordinary daily behavior.

Concerning the independence of the fingers, studies of the frequency of finger usage, again based on observations of ordinary daily behavior, reveal that independent motion of the five fingers is not necessary. The thumb, index, and middle finger are approximately equivalent in importance, while the other two fingers do 50% less than the first three. What is even more important is that human fingers can grasp objects adaptively according to their shapes. This makes it possible for fingers to hold objects more accurately and facilitates the work of the hands even without the fingers moving independently. For this reason, as well as others discussed in connection with prosthetic arms, artificial fingers are modeled on human fingers in the first stage of design.

Therefore, in designing prosthetic arms, most of them are given a five-finger structure and have 1 degree of freedom, as long as they are not function-oriented, such as hook hands. In designing robot hands, there is no requirement that they be similar in shape to human hands, and it is usual to reduce the five fingers to one finger and a thumb and to replace the tips of the hands with other devices according to the purpose for which they are used. There are only a few examples of robots with five fingers, designed for studies based on the analogy to human hands.

MECHANISM

Power Equipment

On the basis of the considerations outlined above, the degrees of freedom are determined. Then the power for each degree of freedom is determined for powered prosthetic arms and robots. The power source is electricity, gas pressure (e.g., CO_2 gas), or oil pressure for powered prosthetic arms, and electricity, air, or oil pressure for robot hands.

Power Equipment for Prosthetic Arms In producing powered prosthetic arms it is necessary to make them light as well as similar in shape to human hands. At present, the arms that have realizable power outputs and are lightest are those in which electronic motors are used as actuators; in many arms cordless motors and batteries (e.g., cadmium batteries) are used for driving power (see Table 1). In this type of arm, prosthetic upper limbs which have only the grasping function have reached the level of commercial realization.

Many of the arms in which gas and oil pressure are used are at the stage of research. These include bellowframe cylinder, piston-cylinder, McKibben artificial muscle, and pouch muscle devices. The former two are used in other equipment as well as prosthetic arms and are not discussed here. The McKibben artificial muscle consists of a rubber inner tube wrapped with spirally wound textile called a sleeve, as shown in Fig. 3(a). The version shown in Fig. 3(b) was developed by Kato et al. and incorporates a textile which suppresses axial extension in the rubber. When air is introduced inside the artificial muscle it tries to expand in the direction perpendicular to the axis and contracts in the other direction, due to the restricted axial motion of the rubber. A twisting motion can also be obtained by changing the pattern of textile in the rubber.

A pouch muscle, as shown in Fig. 4, consists of an accordion-type bag made of nylon film; the device rotates when compressed air is introduced into the bag. A prosthetic arm employing this type

Table 1 Comparison of Power Sources in Prosthetic Arms

Characteristic	Electric power	Gas pressure	Oil pressure
Power source			
Weight	Light	Relatively heavy	Heavy
Maintenance	Simple	Gas cylinder necessary, complex	Fairly complex
Actuator			
Type	Motor	Cylinder rubber artificial muscle pouch actuator	Cylinder
Weight	Heavy	Light	Relatively heavy
Efficiency	Good	Bad	Fairly good
Power	Relatively big	Small	Big
Flexibility	Relatively good	Good	None
Control			
Response	Relatively good	Bad	Good
Wiring and piping	Simple and compact, light	Component relatively big and heavy	Component big and heavy
Control signal processing	Simple	Electric-air conversion sometimes necessary	Electric-oil pressure conversion necessary

of actuator is made by Bussa. In general, gas pressure devices are lighter than electric-powered ones except for the piston-cylinder type, but there are unsolved problems in output power because of the difficulty of high pressurization. Furthermore, because actuators employing this method are generally movable in only one way, achieving one degree of freedom of motion with this type usually requires either a pair of actuators or an actuator used with a combination of springs or rubber parts. There are many devices in which tanks of CO_2 are used as the power source.

An actuator based on oil pressure is not lighter than one using gas pressure, but it can have a higher output-to-weight ratio because higher pressure can be achieved. Therefore, there has been some

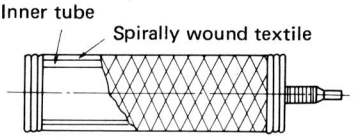

(a) McKibben artificial muscle

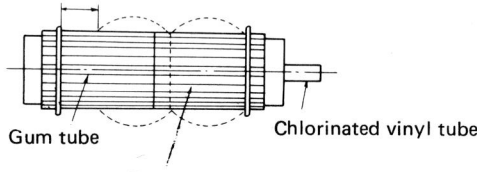

(b) Artificial muscles by Kato et al.

Figure 3 Artificial muscles.

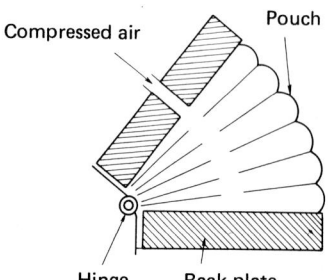

Figure 4 Pouch muscle.

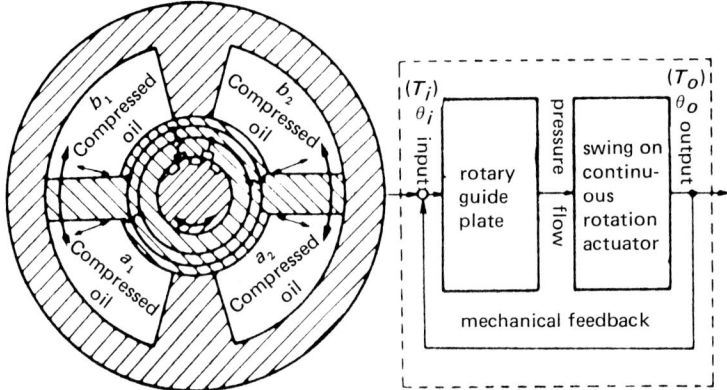

Figure 5 Rotary servo actuator (RSA) made by Mitsubishi metal.

recent research on developing multiple (more than three) degrees of freedom with this method. Cylinders, motors, rotary servo actuators (RSA) and so on have been adopted as actuators. The former two are usually controlled by supermicro control valves (solenoid valves or manual-type valves). On the other hand, the RSA illustrated in Fig. 5, which is a type of torque converter, can achieve similar high output powers when its rotation is driven by the micro motor. It has been the focus of recent research as an actuator for artificial hands. The oil pressure power source usually employs supermicro pumps of the piston or ball piston type driven by DC motors, as shown in Fig. 6. The speed ranges from 3,000 to 10,000 revolutions per minute (rpm).

Power Equipment for Robots For the power units of robots, commercial actuators are generally used, because there are not as many restrictions on weight and geometry as for prosthetic arms. The actuators often employed are AC/DC servomotors and stepping motors for electricity, piston cylinders for air pressure, and pulse motors for oil pressure. With both air and oil pressure there are a few devices that are self-controlled, but in most cases electric signals are used for control. Therefore, in the case of piston-cylinder motors, it is necessary to have some conversion equipment for electric to air or electric to oil, and commercial solenoid valves or servo valves are used for this purpose.

Since industrial robots must be designed for heavy loading, multiple functions, and high precision, most of them are based on the oil pressure method. Many of them use an oil pressure servomechanism in the arm and the open-loop control method with a stopper driven by air pressure in the fingertips. In particular, the electric-oil pressure pulse motor is a combination of electric pulse motor and oil pressure motor with guided valves, and is capable of maintaining a constant relation with input signals through mechanical feedback for the displacement of the output axis because of load fluctuations. Thus it is possible to control position with high accuracy in an open loop, and this type of device is often used to control each degree of freedom in industrial robots.

The electrical method is inferior to the oil pressure method in output power, and so is mainly employed for small, high-accuracy robots. It is used in most artificial intelligence robots because of its ease of handling. The air pressure method is inferior to oil pressure in output power, and is inferior

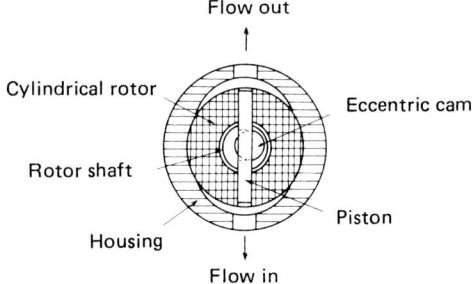

Figure 6 Oil pressure pump for prosthetic arms.

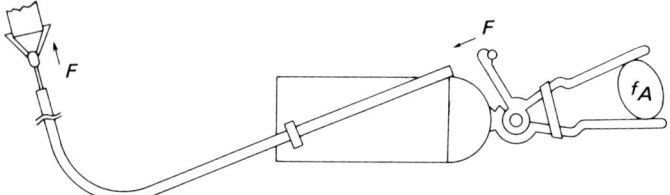

Figure 7 Functional hand.

to both oil pressure and electricity in ease of proportional control, but it is often used in simple robots because of its low cost.

Mechanism

Prosthetic Arms As shown in Fig. 1, there are four types of prosthetic arms. The cosmetic arm has no function other than appearance. The functional arm generally has a hook at the tip of the hand for grasping. The basic mechanism is shown in Fig. 7. A finger of the hook is pulled by a cable operated from a harness at the shoulder and the socket of the hook is pushed by the tip of the amputated limb. The hook is opened by using both forces, and can be closed by a spring incorporated in it when the cable is released. Various shapes of hooks exist, including the well-known Dorance hook, whose tip is bent downward.

In contrast, the tip of powered prosthetic arms generally has a five- or three-finger structure, except for the type developed for thalidomide patients which is a hook. The three-finger structure effectively consists of thumb, index, and middle fingers; however it also has ring and little fingers for appearance in a so-called inner glove, and hence looks like a five-finger structure. Two kinds of fingers are used: the joint type and the single-joint type. In the former case there are the three-joint type of the Belgrade hand and the two-joint type of the Waseda Imasen myoelectric (WIME) hand with the distol interphalangeal (DIP) joint fixed. The single-joint type has all of the DIP joints fixed except for the thumb, and only the metacarpophalangeal (MP) joint movable. The Ottobock hand, which is widely used as a practical hand, is of this type.

Although the weight can be lightened by fixing the joints, the action of the joint type is superior to that of the fixed type. In making the three-point pinch motion the thumb is generally opposed to the index and middle fingers. These are driven by actuators, and there are three types of power transmission for each finger: gear plus link, gear plus cable, and direct drive by gear. As seen in the Belgrade and WIME hands, the output of electric motors is decelerated by gears and the finger joints may be moved by pulling links with ball screws. Usually flexibility is given to the hands by separating the two pairs of index-middle fingers and ring-little fingers and adding springs.

In the gear plus cable method, the fingers are moved by driving pulleys and pulling cables with the decelerated output of the electric motor. Figure 8 shows an example of this type developed in Sweden. The direct drive gear method is employed in the Ottobock hand. The decelerated motor

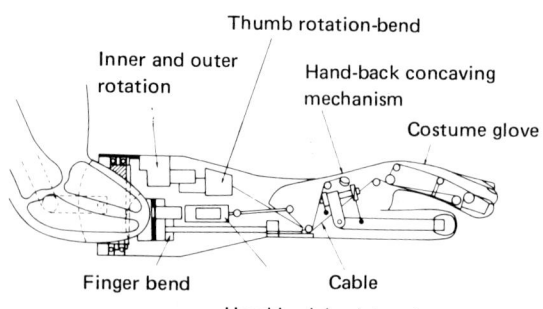

Figure 8 Prosthetic hand made in Sweden.

output can be used for direct rotation of the third joint type. Grasping of objects can be controlled by appropriately adjusting the timing of the motion, since the thumb and the other fingers are located against each other. This can usually be achieved by installing lock mechanisms and/or mechanical flipflop mechanisms in the thumb.

The preceding discussion refers mainly to the electrical method. In the oil pressure and pneumatic methods the velocity is controlled by controlling the fluid flow, so that they do not require the deceleration gears which are indispensable in the electrical method. Apart from this, there are no big differences among the three methods.

Robot Hands Robot hands do not have the geometric restrictions of prosthetic arms and are designed from the viewpoint of the functions they perform. Hence the robot hands developed thus far have more geometric versatility than prosthetic arms.

The robot hands available at present can be classified into three types: (1) hands without fingers, (2) hands without joints, and (3) hands with multiply jointed fingers. The type 1 hands have, instead of fingers, tools that perform their function directly. For example, it is difficult to hold a plate with a hand having a pair of fingers opposed to each other generally, but this can be done effectively by employing pads and a vacuum. Hands of this type are useful for handling fragile objects such as glass plates and bars. Other examples of hands of type 1 are those used as welding guns in spot welding and as spray guns in painting.

Many thpe 2 hands have finger structures—generally the pinching structure with two fingers that can move against each other. Since the objects to be handled are often known beforehand in designing industrial robots, pinching hands have fingertips shaped according to the shape of the objects. There are many hands of this type in which two fingers rotate around a center, and, especially for open-close motion with the two fingers held in parallel, pantograph mechanisms are employed.

This parallel-hold type of hand is generally adopted for the manipulators of deep-sea submersibles, where the objects to be handled are comparatively unclear, master–slave manipulators, and intelligent robots. For the driving mechanism there are types with moving links and direct actuating cylinders, and a combination of motor and rack and pinions. The wire-pulley type is used for master-slave manipulators. A hand having three fingers that can move centripetally can hold round objects and is effective for loading in numerical control (NC) lathes. This type of hand is less common than the two-finger type, but recently several have been reported. So far there have not been any reports of hands with more than four nonjointed fingers.

There are few hands of type 3. Skinner has developed a three-finger structural hand with triple joints driven by motors and links.

SENSORS

In designing artificial hands, it is necessary to provide sensors as well as to determine and realize the degrees of freedom. The touch, pressure, and position sensors can be considered the most important in the hand's perception.

The types of devices used in robot hands and prosthetic arms as sensors are outlined in Table 2.

Table 2 Devices Used as Sensors

Sense	Robot	Prosthetic arm
Touch	Microswitch, electric conductive rubber relay contact, metal conduction	
Pressure	Nozzle flapper, semiconductor gauge, strain gauge, position deviation of servo systems, spring plus potentiometer, plate spring plus differential flange	(Semiconductor gauge) (Strain gauge) (Carbon included sponge)
Position	Potentiometer, encoder	

Chapter 3

Analysis of the Operation of Mechanical Hands

Industrial robots have been regarded as the means of automating various aspects of work, especially the conveying of solid bodies, previously done by human hands. At the same time, robots have been expected to be different from machines whose aim and operating conditions must be rigidly regulated. Therefore, great efforts have been made in development to get rid of such restrictions on automation and to emphasize the universal utility of robots. In spite of these efforts, industrial robots do not completely live up to expectations, and opinions about their universality differ widely between manufacturers and users.

In looking into these technical problems, it is noted that the field of manual work, mainly conveying solid bodies, has been devoid of methods for analyzing operations. Further investigation of individual problems leads to the following observations. With respect to the multiple control of motion of the arm and wrist which characterizes present industrial robots, both the positions and attitudes of the hands can be varied within the work space, but there is little understanding of the role this plays in automation. In addition, the function of hands which is indispensable in the work of conveying solid bodies is separated from the assembly functions of industrial robots. The hand itself is considered one of the accessories, and many hands are specialized ones made exclusively for particular tasks. It is because of this trend that the robots cannot be considered universal, since users find that they must be individually adapted for new tasks.

This chapter discusses the problem of universality as one of the technical problems of industrial robots. In attempting to invent a universal hand one can take the human hand as a model, but with present technology it is not possible to make a mechanical hand with the same functions as a human hand. In this chapter we first discuss the role of fixing solid bodies in the work of conveying them. Then we discuss the functions with which a universal hand should be equipped, analyzing the relation between the restrained factors and the restraining factors. The former are factors related to objective bodies, working environment, working form, and outcome of work. The latter are factors related to the hand. These discussions will give clues to both conquering the technical limitations of the usual specialized hands and promoting the modularization of hands.

ROLE OF HANDS IN AUTOMATION OF MANUAL WORK

In many manual tasks, fixing the object with the hands precedes moving it with the arm and the wrist, so that suitable changes may be made for a particular task. Thus fixing with the hands is directly related to the task, while the arm and the wrist are indirect means of helping to fix and move the object. This is also true in tasks performed by industrial robots.

Because of this close relation between the hands of industrial robots and the characteristics of tasks, solid bodies, and so on, manufacturers cannot freely design the hands. Robots used for handling agricultural products or seafood must grip vegetables or fish, which are soft and easy to damage. In a factory where parts are assembled, various parts are handled. Thus it is easy to understand why manufacturers consider the hands as options. However, as long as the hands are options, robots with specialized hands become specialized working machines, no matter how strongly the universality of the arm might be emphasized.

In recent years there have been efforts to achieve programmable automated assembly; an example is the washing machine gear box simulation by Nevins in Draper. This includes 36 kinds of special hands, jigs, and fixing tools; one hand or fixing tool corresponds to one part, and there are 21 fixing processes in the work. This research has many merits with respect to adaptable control and work scheduling by computer, but it does not seem possible to make modifications in the assembly operation, since it is carried out by many specialized hands.

Thus it appears that automated assembly will be limited, no matter how delicately the robots may be controlled by computer or how many sensors they have, because it is difficult to make the specialized hand parts with multiple shapes and sizes. For this reason, the development of a hand with universal utility is inevitable. A universal hand is needed not only for automated assembly but also for all work involving conveyor systems. There is a modularized hand between the specialized hand and the universal hand.

CLASSIFICATION OF TASKS IN RELATION TO FIXING

The classification of tasks depends on what type of task is thought of as standard. Here, fixing is assumed to be the standard.

Fixing is not only a component of work, but also governs the result. Tasks may be classified in terms of fixing as follows (see Fig. 1):

1 Work that directly changes the attitudes of tools or implements by fixing them: cutting, grinding, welding, painting, testing, sweeping, cleaning.
2 Work in which a partially fixed body is moved: holding, lifting, dropping, contacting, depressing, pulling.

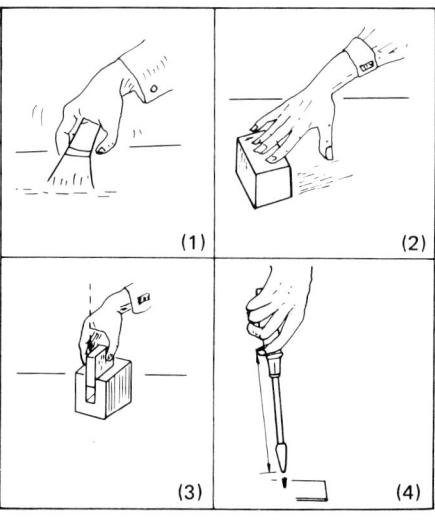

Figure 1 Four kinds of operations classified in terms of constraint.

3 Work in which a body held in a free space is moved: general conveying, carrying in and out, setting, packaging.
4 Work in which a body is indirectly moved through tools or instruments: controlling, steering, screwing, adjusting.

In each succeeding category above, the means of fixing and of control become more complex, increasing the difficulty of automation. Work of the first type can be easily automated because the tools used need not be frequently exchanged, and they can be exchanged without hands. In the second category a hand is required, however.

For the second type of work the hand need only contact the body, and the hand can be a plate or a bar if the contact is stable. In this case, the body should exactly follow the arm or the wrist for the partial fixing; the hand itself must have the capability of complete fixing.

The third type of work is common but presupposes complete fixing of bodies. The hands must overcome the restriction of the surroundings to grip the bodies, and must not prevent the bodies from moving or setting. These problems are taken into account in design, but it is difficult when the bodies, working conditions, and objectives are multiple.

The fourth pattern of work, in which the hand indirectly grips a third body through tools or instruments, mixes the second and third patterns. This type of work requires perception of the attitude or displacement of the body.

Because manual work consists of these patterns and industrial robots are expected to perform the work, universality of hands is desirable.

ANALYSIS OF THE HAND'S OPERATION

Hands have been understood as the means of fixing bodies, but there has been little elucidation of their role in doing this and the factors involved. In this sense, there is little basis at present for the modular design of hands. Therefore the relation between the bodies fixed and the hands doing the fixing must be examined in order to analyze the operation of the hands and their importance. This process may be described as follows:

Identification of the Restrained Factors and the Restraining Factors

Table 1 outlines the results obtained.

Analysis of the Relation between the Restrained and Restraining Factors

This can be done by using the linguistic statement method, the Kawakita Jiro (KJ) method, and so on for causality analysis. Here we adopt the method indicated in Fig. 2 to quantify causality.

In this method a table is first prepared in which the cause factors and effect factors are arranged horizontally and vertically, respectively. Then the influence of each cause factor on each effect factor is investigated. In this case each of the ranks 1, 2, 3, ... is divided into two parts. Although the method of ordering can also be used, the method adopted here is easier to analyze. The same process of ranking is used to construct Table 2 in Fig. 2, where the influence of each effect factor on each cause factor is indicated. Table 3 in Fig. 2 is constructed from these two tables by giving the product of the numbers of each address designated by a column and a row. The sum or the root could also be used, but the product is good for emphasizing the results. If it is assumed that the lower the ranking number, the more closely the effect is related to the cause, then the smaller numbers show stronger causality in Table 3 in Fig. 2.

The method described above has been applied to a causality analysis between the restrained factors (body, environment, and work factors) and the restraining factors (hands and their accessories), with the result shown in Fig. 3. Although the resulting values depend on subjective judgments, a good conclusion is obtained.

Table 1 Restricting Factors in the Functions of Hands

Type of factor	Notation	Restriction	State or condition
Restrained			
Body factors	A	Basic form	Most common geometric form of objects: flat plate, rod, rectangular rod, cube, sphere, etc.
	A	Partial form	Partial geometric form of objects: projection, cut defect, hole, gutter, hollow, etc.
	A	Basic size	Size, thickness, diameter, height, width, length, etc., of standard forms
	A	Weight	Total weight, weight distribution, center of gravity, density
	A	Deformation characteristics	Flexibility, strength, spring coefficient, plasticity
	A	Damage	Fragility, disassemblability
	A	Peculiarity	High accuracy, danger, high and low temperature, expense, magnetism, attachment
	A	Stability	Tendency to slide, rock, or fall
Environment factors	B	Distribution of bodies	Standing alone, multiple series, random loading
	B	State of resting bodies	Plate, gutter, box, hollow, supporting table, hanging
	B	Position or attitude of resting bodies	Instruction, uncertainty, accuracy, direction
	B	Size of work space	Height, width, length, opening angle
	B	Obstacles in work space	Existence of obstacles, transfer and movement of obstacles
	B	Atmosphere	High-low temperature, high humidity, immersion in oil, water, special gas, or electromagnetic field
Work factors	C	Rod size of objects	None
	C	Change of form and size of object	None
	C	Transfer method	Picking up, rotation, sliding, receiving and transferring, conveying
	C	Equipment form	Leave alone, following device, inserting, press-inserting, packing
Restricting			
	E	Restraint form	Semirestriction of pad or receiving, physical restriction (absorb, adhere), mechanical restriction (grip)
	E	Certainty of restraint	Transfer force, stability, accuracy, resistance force, etc.
	E	Covering ability	One-way, dual direction, cross-section covering, cubic covering
	E	Contact state	Soft touch; cubic, plane, or line contacts; number of contacts
	E	Detection and control ability	Visual, touch, pressure, force detection, movement, and force control
	E	Measuring ability	Geometric form, size, angle, temperature, hardness or softness, surface state, slide, accuracy

	Result items	Result items
	A₁ A₂ A₃ A₄ A₅ A₆	A₁ A₂ A₃ A₄ A₅ A₆

Cause items:

	A₁ A₂ A₃ A₄ A₅ A₆
B₁	1
B₂	3
B₃	2
B₄	1
B₅	2
B₆	3

table 1

table 2: 1 2 1 3 2 3

	A₁	A₂	A₃	A₄	A₅	A₆	Σ/6
B₁	1 ×	1 ×	2 ×	1 ×	3 ×	2 ×	3
B₂	3 ×						
B₃	2 ×						
B₄	1 ×						
B₅	2 ×		table 3				
B₆	3 ×						
Σ/6							

Figure 2 Tables for causality analysis.

	E₁	E₂	E₃	E₄	E₅	E₆	Σ/6	2 or less	3～4	5 or more
A₁	2	6	1	4	2	3	3	3	2	1
A₂	2	6	2	2	2	3	2.83	4	1	2
A₃	6	9	2	9	2	1	4.83	3	0	3
A₄	2	1	2	9	6	9	4.83	3	0	3
A₅	4	1	6	2	2	6	3.5	3	1	2
A₆	4	2	6	1	1	9	3.83	3	1	2
A₇	2	1	6	3	3	9	4	2	2	2
A₈	6	4	1	2	3	6	3.76	2	2	2
B₁	6	9	2	6	1	3	4.5	2	1	3
B₂	4	9	1	4	2	9	4.83	2	2	2
B₃	6	9	9	2	1	2	4.83	3	0	3
B₄	3	9	6	6	3	6	5.5	0	2	4
B₅	9	4	2	4	2	9	5	2	2	2
B₆	1	4	6	3	9	6	4.83	1	2	3
C₁	3	9	6	6	3	9	6	0	2	4
C₂	4	6	3	4	2	3	3.67	1	4	1
C₃	1	3	1	2	2	6	2.5	4	1	1
C₄	4	3	1	2	2	6	3	3	2	1
Σ/18	3.2	5.3	3.5	3.3	2.77	5.36	4			
2 or less	6	4	10	7	12	2				
3～4	7	5	1	6	4	4				
5 or more	5	9	7	5	2	12				

Figure 3 Correlation table among constrained and constraining factors.

According to this analysis, the restrained factors which have strong causality are the transfer form of bodies (C_3), the special (partial) shape of bodies (A_2), the basic shape of bodies (A_1), and the setting form of bodies (C_4). Restrained factors with weak causality are the rod size of the objects (C_1), size of the work space (B_1), and obstacles in the work space (B_5). The tendency of the result is common except for the rod size, but this factor is not prominent in the analysis of fixing because it is important for specialized machines. The meaning of the factors that have strong causality should be fully understood in the design of hands.

Second, the restraining factors which are found to have strong causality are the ability to detect and control (E_5), the form of fixing used (E_1), and the covering ability (E_3). In contrast, the measuring ability (E_6) and stability of fixing (E_2) have weak causality. E_5, the ability to detect and control, is one of the factors used directly in fixing, and it is related not only to the mechanical hand itself but also shows the importance of sensors. E_1 and E_3 are of general importance. E_6, the measuring ability, is relatively unimportant because it is not required for fixing. Of course, one way to design sensors is to base them on the ability to measure, but it is an additional condition independent of fixing. Finally, the weak causality of E_2 can be understood, since this factor is not directly involved in fixing but is the result of fixing.

When the method outlined above is applied to actual operations, each item should be weighed after it has been determined which of the restrained factors are important in the work. Then one can get rid of factors that are unnecessary and add others. In this case, the balance between factors should be taken into account. The restraining factors obtained by applying this method largely determine the function of the mechanical hand.

Representation of the Operation of Mechanical Hands

The results of the causality analysis apply not only to mechanical hands but to conveying systems in general. The surrounding factors, even if there are not many, are introduced in the form of the arrangement of bodies and the obstacles in the work area. Then the restraining factors are abstracted from the results for mechanical hands, and it is determined how the functions should be set.

Function of Detection and Control In the preceding discussion, this function is used to detect the circumstances of the objects or the surroundings that must be taken into account for fixing, and then, based on the results, to control all movements in order to prepare the conditions for fixing. The detection involves sensors used for seeing, touching, pressing, and strength. The control involves the wrist and the arm as well as the hand. The ability to detect and control objects has been examined in studies by Thuboi (Mitsubishi Electric Company) and Tani (Mechanical Engineering Laboratory), where hands have been provided with sensors for visual detection.

If sensing and control are limited to the operation of fixing, the objects must be in the hand's work area and it must be possible to sense the difference between the objects and the surroundings. Then the contact with the objects and the force applied to them can be controlled only for aimed fixing.

Fixing Function This function involves the fixing form, contact, and covering ability, and it can be divided into two categories: partial fixing (holding, contacting) and complete fixing.

In determining the functions of the hands in fixing bodies, the free body can be thought of as moving in six directions along three axes, X, Y, and Z, as illustrated in Fig. 4. Thus the transfer movement has 6 degrees of freedom. In addition, the body can rotate in six directions in the XY, XZ, and YZ planes, giving 6 degrees of freedom of rotation. Partial fixing involving some degrees of freedom can lead to complete fixing by using the restraints of gravity and the surroundings. In complete fixing the hand has 12 degrees of freedom, but it is common to neglect the rotational degrees of freedom.

The operation of fixing a body with the hand involves the hand itself, the body, the contact, and the covering surface on the contact part. Fixing by the contact part is illustrated with the model in Fig. 5. Figure 5(*a*) shows a point contact, which fixes 1 degree of freedom in the direction of the contact. A one-point contact cannot fix the remaining 11 degrees of freedom. Figure 5(*b*) shows a

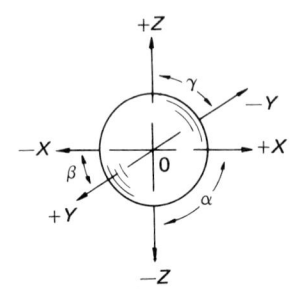

Figure 4 Motions of a body.

line contact, which fixes 7 degrees of freedom. A surface contact, shown in Fig. 5(c), can fix 11 degrees of freedom. However, the line and surface contacts involve 4 and 6 degrees of freedom related to friction, respectively, and these components depend on the relation between the friction coefficient and the ratio of tangential force to normal force. The surface contact achieves holding by gravitation and contacting by external force.

The type of contact is restricted by both the shape of the body and the contact surface of the hands, and so is determined not only by the hands. However, a surface contact can be readily made if the contacting surfaces of the hands are made of deformable materials. This enhances the ability of the hands to fix objects in spite of their shape by increasing the contact surface.

Physical Fixing and Mechanical Fixing A body is fixed physically or mechanically depending on the form of the force applied by the hands. Physical fixing usually involves use of a vacuum or electromagnetic force, and there is one contact surface. Airtightness of the body is required for use of a vacuum, and magnetism is necessary for the electromagnetic method. Mechanical force is used in mechanical fixing, and there are three types of fixing corresponding to the degree of mechanical covering of the body: pinching (face to face), gripping (surface covering), and holding (body covering). The type used generally depends on the shape of the body, that is, pinching is used with plates, gripping with rods, and holding with lump materials.

Interference with the Working Environment As the degree of contact space or covering goes up, interference with the working environment increases, although the ability to fix bodies may also increase. Environmental constraints become more severe in the order of absorbing, pinching, gripping, and holding. This problem must be taken into account when the degree of contact space is enlarged—for example, as Landstrom of Sweden has done in order to increase the fixing ability of vacuum absorption.

Soft Touching Soft touching by the contact part enlarges the area of contact and decreases the pressure that must be used. With this type of contact part it is possible to hold fragile bodies and to absorb external shocks by use of springs in the mechanism. It is also possible to set bodies in accordance with the working environment (for example, the Fanac hand). The fixing forces and external forces can be determined by measuring the displacement of cushions.

Control of the Fixing Force In most mechanical hands the fixing force is supplied by a vacuum, oil or air pressure, or motors. One method has been patented in which the fixing force comes from a spring, which is opened by contacting the surface. Hands based on this method could be called nonenergized. In the other cases, the ability to control the force used limits the maximum weight and

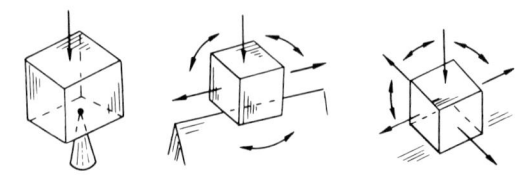

(a) Point contact, (b) line contact, (c) face contact   **Figure 5** Types of contacts used in fixing an object.

determines whether the hands can fix a soft or a brittle body. When a hand can freely select among rigid fixing, elastic fixing, and constant-force fixing, its adaptability to the work is higher.

Holding with the Centerline Fixed and Measuring In present industrial robots the arm and the wrist are used to control position and attitude, and a hand that can fix various bodies with a constant attitude has been eagerly sought. In holding a rod, the hand is expected to hold the centerline constant. There are several hands that can synchronize three nails in a peripheral direction. The function of measuring is mainly used not for fixing, but for determining the diameter or the width of rods.

PROBLEMS IN DESIGNING MODULAR HANDS

A mechanical hand may be considered to have seven functions, and the modular design of hands will be realized when these functions are regarded as standard. When specialized hands can be replaced by modular hands, industrial robots will be universally used to automate work involving conveyor systems. Then programmable automated assembly will become possible, and complete factory automation will no longer be impossible. The problem in achieving this does not always lie in designing the force function modules for hands. It is first necessary, for modular hands to be used in different fields and in a variety of applications, to closely analyze the attitude of bodies on conveyors and the tasks done in actual factories, as has been done in Europe, particularly Germany. The next step is to design modular hands for use in different fields. The results of this step will form a general groundwork for modular design.

REFERENCES

1 J. L. Nevins and D. E. Whitney, 4th International Symposium on Industrial Robots paper, pp. 387–406, 1951.
2 Y. Tsuboi, 5th International Symposium on Industrial Robots preprint B3-21-32, 1951.
3 K. Tani, 4th Robot Symposium preprint (in Japanese).
4 G. Lundstrom, 4th International Symposium on Industrial Robots preprint E3-25-36, 1951.
5 H. Ito, *Biomechanism 3,* edited by the Biomechanism Society, pp. 145–154, 1975.

Chapter 4

Grasping in Handwork

The word "handwork" is often used in daily life. Also, in connection with the automation of technology, production, and computational systems, a positive attitude toward mechanical and automated handwork has recently been observed. However, discussions of the characteristics of handwork and the reasons why its automation has not been easy have not been very enlightening. In this chapter we discuss handwork and its mechanization and automation by considering the problem of grasping.

RESTRAINT OF THE OBJECT OF HANDWORK

In general, handwork does not refer only to work done with the hands, but also involves a wide range of human operations including movements of the arms, waist, and legs. Therefore the work is not defined only in terms of the parts of the body involved. How handwork is defined depends on what one means by "work" and also by "hand."

Handwork is the process involved in changing material from the object state to the product state. For this process to take place it is first necessary to place the object within the worker's sphere of operation and to restrain it. Next, power is used to make changes in the condition of the object. The worker's role at this stage is to observe the condition of the object and to control the movements needed to make changes in this condition to convert the object to the product state. The movements involved in making these changes to attain the product state are not necessarily identical, hence human handwork must be flexible.

In any case, the first step in practicing handwork is to restrain the object, and the selection of the next action is based on the observation of the change in condition due to restraining the object. Because many human operations involve use of the hands as restraining devices, we employ the term handwork.

To summarize, handwork makes changes in the condition of objects by restraining them by hand. In the relation between work and action, the means used to fulfill the restraining condition is considered very important. The hands are generally used for restraint. Movements of the arms, waist, and legs are supporting actions and are indirect means of changing the conditions of objects.

HAND RESTRAINT AND GRASP

We consider handwork as effecting a change in condition. If the purposes or conditions of the work are changed, even though the action against the same object remains the same, the meaning of hand restraint changes. That is, the worker determines what restraint of the object is needed in order to achieve the purpose of the work and determines the movement of the hand accordingly, rather than restraining the object by moving each part of the hand structure. Any movement can be adapted to achieve the restraining condition needed for the purpose of the work, and even though the movements of hand structures are limited, it is possible to achieve virtually unlimited actions. However, mechanical hands cannot imitate the movements of human hands. The movements of human hands fit the entire body structure, while those of mechanical hands must fit the mechanical structure. In both cases, restraint must be used to accomplish the purpose of the work.

Humans learn through experience to understand the relation between the actions taken and the purpose of the work. Differences in actions between skilled and unskilled workers are related to their degree of understanding of the work and how accurately they can relate their actions to the purpose and conditions of the work. This understanding cannot be acquired only through theory but requires learning through experience.

We engineers have not attacked these problems logically, and therefore do not know how to design machines that can respond like humans to the purpose and the conditions of work, even though we can perceptively suggest work schedules or design machines for certain conditions. It is important to understand how humans work in order to grasp the structure of work in a logical way. This will be the starting point for a broader range of automation.

From the working forms of humans we have extracted the restraining action of grasping as the action that characterizes handwork. Many other restraining actions besides grasping are involved in human handwork. Before classifying these actions, however, let us consider the general meaning of restraint of objects in handwork.

Suppose an object is in three-dimensional space and has the capability of moving around that space freely. Then, as shown in Fig. 1, movements can be broken down into movements along X, Y, and Z axes. Considering two directions (+ and −), there are six movements. Also, if the object is rotated around the X, Y, and Z axes, there are 6 degrees of freedom of rotation.

The restraining movements of the hand can be divided into several categories according to the degree of restraint in the 12 degrees of freedom of an object. The degree of restraint is determined by the purpose of the work. Generally, the action of grasping is used to restrain the movement of the object completely by taking it into the palm and structurally covering it.

Since humans often use parts of the structure of the working environment to aid in restraining an object, hand restraint of objects by humans is often incomplete (less than 12 degrees of freedom). The action of "pressing" is a good example of partial hand restraint. Also, the force of gravity on an object is used in the action of "hanging."

Thus humans can achieve complete hand restraint of an object by using structural, body, and friction forces and other conditions in the working environment, or they can restrain the object incompletely, making it possible to replace or rotate it.

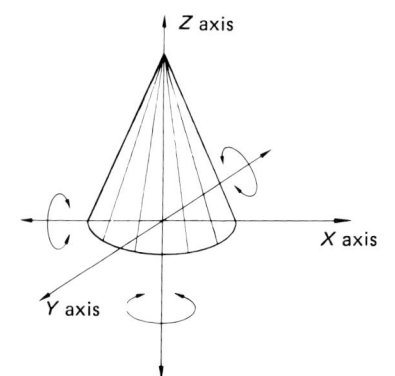

Figure 1 Motion of a body with 6 degrees of freedom of transfer and 6 degrees of freedom of rotation.

Present mechanical techniques use "attaching," "binding," "connecting," or "absorbing" methods in addition to mechanical restrictions, and they also widely use restraint by physical power. The absorbing contact method is often used in production since it is easy and convenient to use vacuum and magnetic absorption contacts; however, there is a limit as far as the strength of the restraint is concerned, and use of the method depends on the working conditions and characteristics of the object. In contrast, the use of grasp is exceptional, and it is a good reference action for a hand with a wide range of applications because of its broad adaptability to different working conditions and objects.

MEANING OF GRASPING

In grasping an object, the human hand restrains it completely with body and friction forces by moving the fingers and palm and covering the object positively. Thus the characteristic of the human grasp is the tendency to cover an object. As shown in Fig. 2, restraint by grasping can be divided into three categories: pinch (or pick-up type), grip, and nip (or snap type).

The action in "pinch" restraint is employed to cover objects, and is used for objects with three-dimensional shapes such as balls and for small objects that fit the length of the hand structure. "Grip" restraint is used for stick-shaped objects and is a two-dimensional covering action; to use this type of restraint it is necessary to recognize the direction of the axis of the object beforehand. In "nip" restraint the object is placed between the restraining devices; generally this is used for objects such as boards. These means of restraint are illustrated in terms of the grasp of the human hand, but they have been generalized for machines. For example, the mechanical version of pinching is achieved with a bucket, gripping with a chuck (in lathes), and nipping with a vise.

It may not seem necessary to discuss any of the factors involved in restraint; however, the relation between these descriptive names and the restraint function is often not understood correctly. Grasping is understood as the way to restrain objects. The substantial difference between grasping and any other means of restraint is that grasping actively removes the object's environmental restraints rather than receiving the object under the controlled conditions of the workplace and then restraining it. The characteristic of human hands is that they are able to choose the grasping action used according to many different working objectives.

FUNCTION OF GRASPING

Currently, industrial robots are equipped with optional hands, designed and made for specialized uses depending on the purpose and conditions of the work. As long as this type of hand is used, it is not

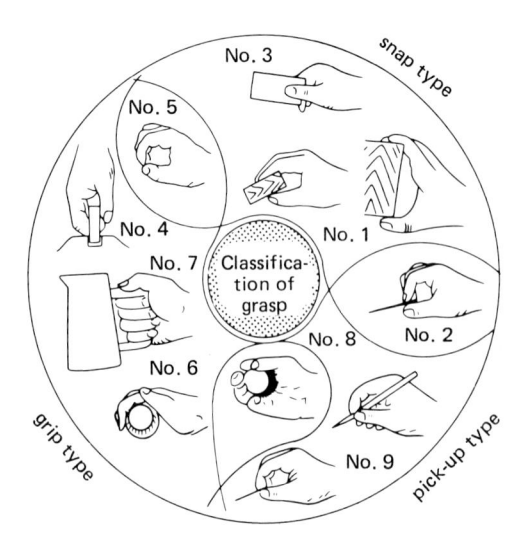

Figure 2 Grasping with the human hand.

necessary to tackle the problems of grasping, since specialists will handle the problems for particular cases.

Here we are concerned with the functions of a universal hand. Before discussing the function of grasping with a universal hand, however, let us consider briefly how the universality of the grasp of a human hand can be related to the grasping function of a mechanical hand.

Human hands can draw water, give finger signals, press, grasp, and hold; they can also perform compound actions such as changing grasp. The ideal situation would be to have mechanical hands emulate these actions.

However, mechanical hands have not yet reached this stage, and what we should think about is the most effective action for mechanical hands to adapt functionally. In order to lessen the functions of mechanical hands, we must divide the functions of human hands into those that can be adapted to mechanical hands, and the functions of tools and machines into those that can be handled by mechanical hands. In particular, an industrial hand is made only for handling tools and machines for a set purpose, not for drawing water or writing a letter. A cup can be used for drawing water and a typewriter for writing a letter. The actions involved in the latter two activities can be summarized as "gripping" and "hitting"; the hitting action does not necessarily require the function of a hand and is often accomplished by using sticks to express the movements of wrists and arms.

Figure 3 outlines the functions of the hand. The function of the human hand that must be adapted to the mechanical hand is grasping. If mechanical hands can perform this function widely, as human hands do, they will be able to work very similarly to human hands in the range of single-function tools and machines which they can operate. Therefore we consider human hands in terms of the grasping function.

As mentioned above, grasping is used to restrain objects in various forms depending on the purpose of the work. At the same time, the hand must have the added ability to adjust to the working environment and be able to grasp and restrain objects under various working conditions. Adjustability to different conditions may be used as a measure in evaluating the generality of hands.

The problems discussed above involve not only the hand mechanism but also the ability to recognize the conditions of the work and the object and the ability to react to these conditions and control movements. These abilities apply not only to grasping but also to other actions necessary in work. It seems clear that recognition ability is central to achieving a universal hand; however, we must find out how to deal with the problems involved.

In considering the current status of mechanical hands, it is important to determine the extent to which universality can be achieved with input information and controlling instructions. For the past five years I have examined the structure of the human hand and its ability to control movements in the grasping function, and tried to simplify these aspects of the human hand so that they can be applied to the objective of universality. As a result, I have found that imitating the structure of the human hand is not a premise for designing a mechanical hand with the grasping function. Several examples bear this out and are discussed in the following section.

ANALYSIS OF THE GRASPING FUNCTION OF HUMAN HANDS

The first research carried out involved the structure of the fingers of the human hand. Their effect in restraining objects was examined theoretically by considering fingers with three types of joint structure: a free-joint finger (finger with one joint), a fixed-joint finger (finger with a fixed curve), and a no-joint finger (straight finger). It was found that a multiply jointed structure like the human finger was not absolutely necessary for a hand capable of restraining objects of five basic shapes (ball, round stick, square stick, cone, and flat plate). It was also found that a hand consisting of opposing fingers (a fixed-joint finger and a no-joint finger) connected by a joint was good enough. This type of hand structure uses pinch-type restraint for three-dimensional objects (ball, cone), grip-type restraint for two-dimensional objects (round stick, square stick), and nip-type restraint for one-dimensional objects (flat plate) (see Fig. 4). However, this hand structure can grasp objects only if they are floating in the air, and it is not discussed further in connection with restraint.

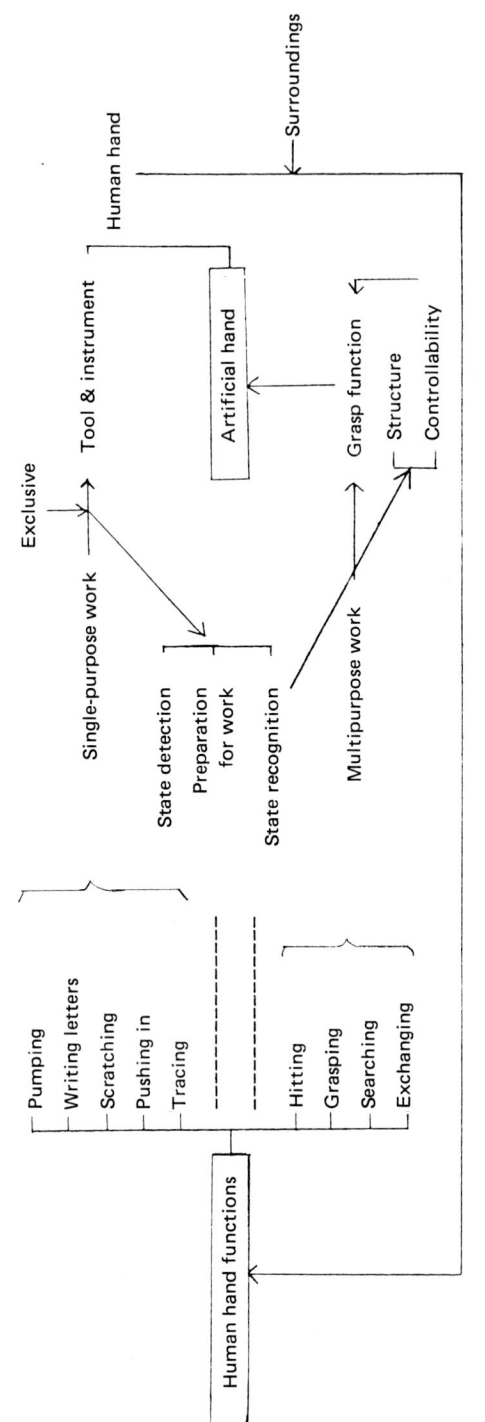

Figure 3 Functional roles of an artificial hand.

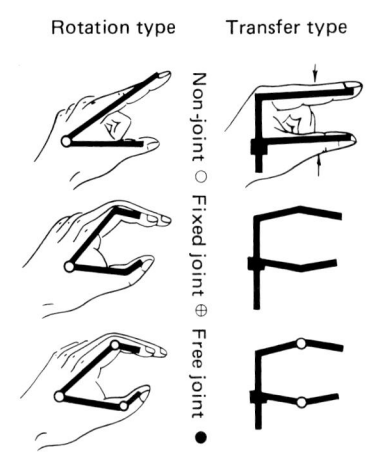

GRASPING IN HANDWORK

Figure 4 Model of the human hand and a mechanical hand.

Next, the effect on grasping of the number of fingers and joints and their structural arrangement was examined in relation to the conditions of the working environment and covering the object. It was found that as long as the thumb is opposed to the other fingers and the palm, a three-fingered structure is sufficient. For a nine-jointed structure with coordinated movement of the joints, it becomes more difficult to cope with the working environment as the covering of the object progresses through pinching, gripping, and nipping. Also, a flat plate on a table can be grasped with the human hand by raising it with three fingers (six joints) only when the fingers have arbitrary joint movement. This kind of routine grasping movement requires a high degree of recognition and is limited by the finger structure. Also, the softness of the skin of human fingers has the same effect as joints in covering three-dimensional objects (e.g., balls), and such objects can be grasped by pincers or tongs if sponges are pasted on their pinching edges.

Finally, an experiment conducted four years ago showed that the bending movement of the finger with three joints did not control grasping on a very high level if a special raising movement involved in grasping was not performed.

For general grasping, three joints can be adapted to three different patterns of movement, depending on the external power applied to the fingers. These patterns are:

1 Three joints moving in regular relation.
2 Basic joint bending independent of the other two joints.
3 Two joints bending while the basic joint rotates.

The third pattern of movement has a particularly important function in bringing the fingers close to the object to follow the surface of the work. As shown in Figs. 5 and 6, the bending movements of the fingers mentioned above are imitated by the control of a link mechanism, which makes it possible to develop a type of universal hand called an adaptive shape hand.

Efforts to mechanize and automate human handwork are expected to aid in the design of prosthetic arms for upper-limb amputees. The problems involved in this require an understanding of handwork, and we are only at the beginning stage. A great effort must be made to understand the grasping function of the hand in order to develop a universal mechanical hand.

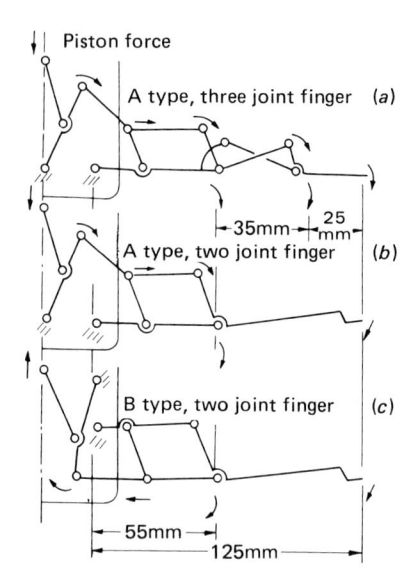

Figure 5 Mechanism of action of an adaptable hand.

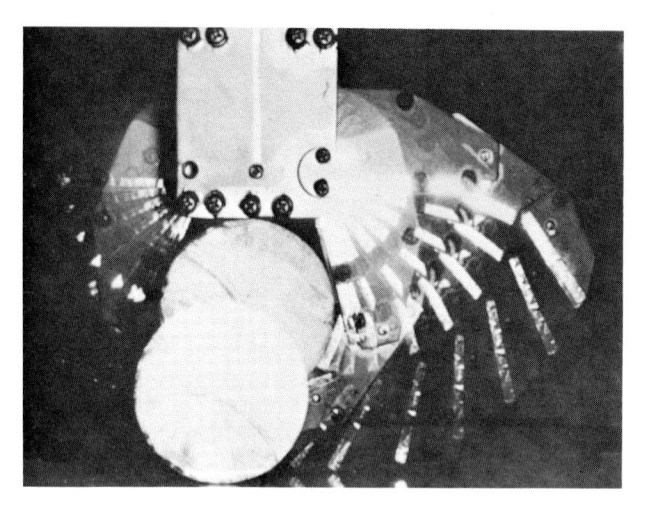

Figure 6 Grasping action of a mechanical hand.

Chapter 5

Characteristics and Design of Power Sources

In designing power sources for robots and mechanical hands, an important factor is whether the robot is self-moving or fixed. For a human-type robot capable of walking around, it is desirable to minimize the weight of the energy storage devices (battery and gas cylinder) and the conversion equipment used to convert the stored energy into power (electric power, oil pressure, air pressure, etc.). In contrast, for ordinary industrial robots, the most important considerations are adjusting the movement of the robot to its requirements and building the robot in an inexpensive way.

Power sources must be chosen according to the form and purpose of the robot. The general structure of a power system for robots is shown in Fig. 1. It consists basically of three parts: an energy storage device and/or generator, conversion equipment to convert the energy (from the storage device or generator) into power designed for the robot actuator and transmitted through the power transmitting medium, and an actuator to supply power to the robot.

One difference between the power systems of self-moving and fixed-type robots is that the

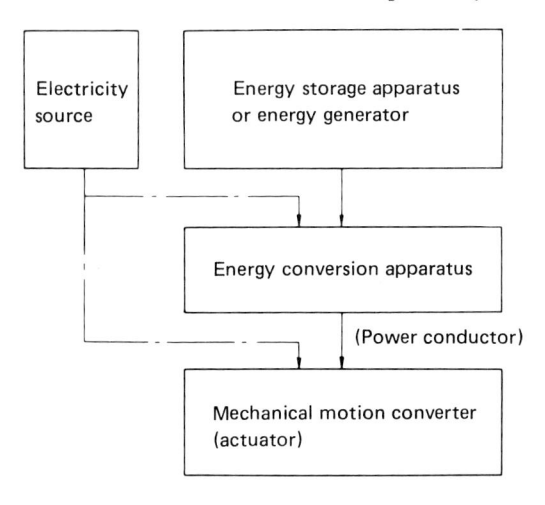

Figure 1 Structure of the power system for a robot.

self-moving robot must be built with energy storage devices or generators, while the fixed-type robot can use an electric power source or an external compressed air source. However, the energy conversion equipment and actuator can have the same structure for both types of robots. As shown in Fig. 2, electricity, oil pressure, air pressure, and other mechanical methods are generally used. At the present level of technology, these four methods are the only utilizable power sources.

CHARACTERISTICS OF REPRESENTATIVE POWER SOURCES

In choosing the optimal design for the power source, the following factors must be considered:

1 Operating power (maximum power the robot can produce, e.g., maximum lift force)
2 Response speed (maximum operating speed)
3 Size and weight (limiting size and weight of the robot)
4 Safety (protection against heat and overloading)
5 Ease of use
6 Life span
7 Cost and operating cost

Operating power and response are the most basic capabilities. When these have been determined, the power source is chosen by taking into account size and weight limits, safety, ease of use, life span, cost, and operating cost. However, for self-moving robots the weight of the power source is often the most important consideration. For industrial robots, ease in handling and safety are the important factors. Thus the ranking of items on the list above must depend on the type of robot.

Table 1 shows a comparison of the electric, oil pressure, and air pressure methods. Apart from its advantages in cost and safety, the air pressure method is inferior to both the electric and oil pressure methods. The oil pressure method is superior to electricity in maneuverability, response speed, weight, and size, while electricity is superior to oil pressure in ease of use, cost, and maintenance. However, the electric method becomes superior if an energy storage device and generator are included. The results of these comparisons also differ depending on the size, type, and purpose of the robot.

Electric Power Method

For information and signal processing, methods based on electricity are unequaled. Power sources and transmission line networks are available almost everywhere in the world, and electrical energy can be stored in batteries. Thus the biggest advantage of the electric power method is that it can easily be used to form many kinds of systems. In particular, if power sources such as engines are not available, the energy supply must depend on electricity. Since it is not necessary to convert electricity into another form of energy, this leads to advantages in cost, size, and weight.

For the actuator, for uses requiring less than medium power, the electric servo method is often used. The oil pressure method has been comparable to the electric power method, as shown by the history of competition between electric servo and hydraulic servo methods for machine tools. As far

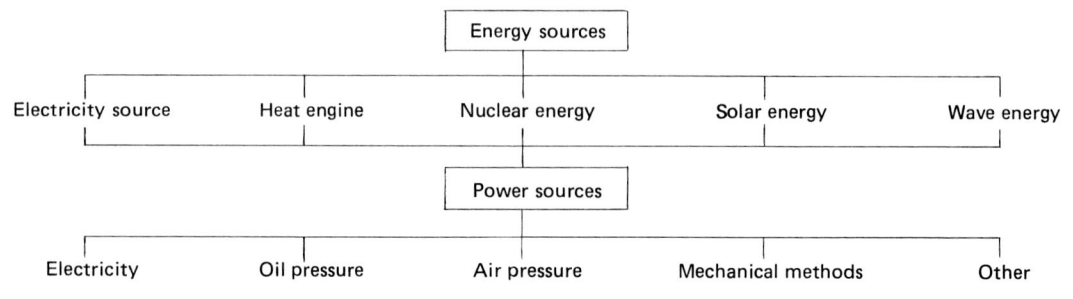

Figure 2 Types of power sources for robots.

Table 1 Comparison of Various Power Sources

Factor	Electric method	Oil pressure method	Air pressure method
Maneuverability	Low to medium power obtained, normally rotational force	Very high power obtainable, rotational and rectilinear forces	High power not available, normally rectilinear force
Response speed	Improved by the development of low-inertia servomotors, close to the oil pressure method in low and medium power outputs	High; large torque-inertia ratio makes rapid response possible	Generally low, difficult to obtain high-speed response; because of the smaller pressure drop, it is possible to obtain a quicker response for a simple movement than with oil pressure
Size and weight	Improved by print motors, wide range of selection possible	High ratio of weight or size to output power, power unit dominates space	Inferior to the oil pressure method; more valuable for low outputs
Safety	Susceptible to overloading, anti-explosion safeguards necessary, otherwise safety rated high	Heat generation, danger of fire	Least susceptible to overloading, no heat generation, safe for human bodies
Ease of use	Peripheral devices complete, inspections easy	Control of working oil, pipe flushing, filtering necessary	Necessary to remove water in supply air and to add lubricant, but easier than oil pressure method
Life span	Improved by use of solid-state thyristors	Long, aided by the oil lubricant	Inferior to oil pressure and electricity because of lack of lubricant in the air
Cost	Average	Price and running cost both high	Low

as machine tools are concerned, the hydromatic power servo was once considered absolutely superior because of the rapid progress made in improving the efficiency and ease of handling of electronic-hydromatic servo valves. However, as a result of the development of quick-response motors such as the low-inertia servomotor and thyristor amplifier, electric power servo has greatly progressed; in addition, its simplicity and ease of handling have led to electric power servo regaining its predominance for less than medium-power applications.

Size and weight have been reduced as servomotors have progressed, so that electric power servos can now be the smallest devices for low power, although their power-to-weight ratio is inferior to that of hydromatic servos for medium and high power. Also, electric power servos have been improved in life span, maintenance, and safety, except for safety against explosion.

The disadvantages of the electric power method are that it is susceptible to fire or to breakdowns due to heat caused by overloading; rotational motion is readily achieved but rectilinear motion requires the use of other mechanical devices (the recent development of linear motors is solving this problem); and in situations where explosions are possible, strong safeguards against explosion are needed.

Oil Pressure Method

Oil pressure or hydromatic power sources ordinarily require a hydromatic power unit to convert electrical energy into a high-pressure flow of oil. For this reason, the hydromatic power method involves one more piece of equipment—the power unit—than the electric power method.

However, hydromatic power sources are easily able to produce very high pressures, so that comparatively small equipment can provide large amounts of power. The main characteristic of hydromatic power methods is that they can be used to obtain high operating power easily. For aircraft and construction vehicles, which have comparatively large power generators, hydromatic power units are much smaller than electric motors but can produce high power outputs. Therefore, hydromatic units are considered excllent power sources.

From the viewpoint of the actuator, the characteristics of hydromatic systems are:

1 Very superior ratio of output power to size or weight of equipment. (However, for a whole power system hydromatic power units are usually needed, so this is sometimes inferior to other methods.)

2 Excellent speed of response. As high output powers can be obtained from small equipment, the torque/inertia ratio is large, and very quick response can be obtained. It is also possible to control the motion over a wide range with high accuracy.

3 Long life span. Since oil is the working fluid, lubrication is good and metal corrosion is low.

4 Frequently, a larger floor space requirement for the equipment in factories. It is often necessary to build an independent hydromatic power unit into each robot because common use of a hydromatic power source leads to reciprocal interactions. Hence more space is needed for the equipment.

5 Susceptibility to dust in the working oil. Most accidents with oil pressure devices are caused by dust in the working oil, and repairing breakdowns in the devices takes a relatively long time. Hence it is very important to keep the working oil clean.

6 Relatively high heat generation, serious stains due to leaking oil, and noise are problems. Disposal of waste oils is becoming a local problem.

Air Pressure Method

The usual power source for a pneumatic power device is an air compressor. Since air has high compressibility, a common air pressure source can be used without much reciprocal interaction, so that the floor space requirement is smaller than with the hydromatic power method.

Energy storage is possible, since even a small compressor can store enough pressure in a tank to produce high momentary power. It is possible to use a pressure storage cylinder like a battery by accumulating highly compressed air or gas in the cylinder.

Although this method is currently used only in missiles, it is possible to use highly compressed gas as a power source by burning it as a fuel. If the problem of heat could be resolved, it would be possible in this way to develop very small, portable power sources.

Usually, with pneumatic power devices the usable pressure is only 5 to 7 bars. This has the advantage of safety, but the disadvantage that high operating power cannot be obtained.

Another disadvantage of pneumatic power devices is that they require lubrication with an oil mist or some other method because of the lack of lubricants in the air. Also, because of the high compressibility, the system has low efficiency. However, this can be improved for the whole system by energy accumulation, such as by the installation of tanks.

Since water in the air causes machinery to rust, it is necessary to dry the supply air. Oil mists used for lubrication lead to pollution of the air, and exhaust gas and noise are also problems.

Other characteristics related to the operation of a robot powered by a pneumatic device are the following:

1 Operating speed and output power can be controlled arbitrarily and easily. However, accurate control is difficult because friction will increase with compressibility and working pressure.

2 Pressure drop in the pipes is low, hence a high-speed response can be obtained.

3 Because of the compressibility, it is difficult to stop and hold in the middle of an operation with accurate control of speed and position.

4 Centralized control is easy as it does not need a return point. Changes in temperature are not a problem, and this method can be used over a very wide temperature range and especially good for use in high-temperature areas.

5 Excluding efficiency, the cost is low.

Pneumatic power devices are suitable for light operation and for low-cost applications which do not require high-level control. A compressed gas power source has the potential of being an excellent power source for self-moving robots. Also, gas cylinders are superior power sources for prosthetic arms and legs.

Other Power Sources

There are other power sources besides those discussed above, but they are used very little. For example, mechanical energy can be stored by use of a flywheel and heat energy can be used directly, but there are still technical problems with these methods. However, concern about energy conservation and the wish to switch from oil to new sources of energy make the development of new and effective power sources desirable. Among the methods considered are the use of wave power to generate electricity, the use of solar heat in solar batteries, and the use of nuclear energy. For the actuator, a mechanical power system has been considered. In this method a revolving movement is produced by a motor, using a clutch and cam. This is an inexpensive, quick, and accurate method and is widely used in automatic machines; however, it produces only one type of movement and it is difficult to program its operation.

POWER SOURCES FOR SELF-MOVING ROBOTS

A self-moving robot carries its own power source in its body. For this reason it is necessary to minimize the ratio of weight to output power. Usually, the power source used is either driven by a motor, like an automobile engine, or based on electrical energy stored in a battery. However, these methods can be used only for robots with wheels; they are not suitable for biological-type robots or prosthetic arms and legs. For the latter uses it is necessary to choose the lightest power source and accept a compromise in the efficiency obtainable at the present level of technology.

Table 2 shows the currently available energy density per unit weight of portable power sources.

Table 2 Energy Density per Unit Weight for Energy Storage Equipment

Type	Example	Energy density per unit weight (Wh/kg)	Comment
First stage battery	Mangan dry battery	20–80	
	Air wet battery	120–210	
	Lithium battery	250–480	
Second stage battery	Lead (Pb) battery	20–40	
	Alkali battery	20–40	
	Silver, zinc battery	60–110	A type of alkali battery
Fuel battery	Alcohol system	70–80	
	Hydrogen system	300–350	
Physical battery	Silicon solar battery		
Flywheel	PRD-49 filament	422	Compound material
	Boron filament	300	Compound material
	Steel flywheel	48	
(Liquid fuel)	Gasoline	20,000	Net
		5,000	Mechanical efficiency 25% (average)

For first-stage batteries, the energy density per unit weight can be higher than 300 Wh/kg with lithium batteries. The fuel battery has excellent characteristics, but its high cost and short life span are problems.

Although it has become possible to obtain high energy densities with batteries, they are still inferior to heat engines using gasoline as a fuel. Hence it is desirable to develop extremely small heat engines which are portable and protected with gas against high temperatures, like the small highly compressed gas generators developed for satellites or missiles.

In addition to batteries, compressed gas cylinders are used as energy storage equipment for prosthetic arms. Although they are small, these cylinders produce relatively high momentary power, and this method is comparable to the electric power method. It is difficult to compare the two methods with respect to energy density per unit weight. The difference varies in different situations, and in the case of the compressed gas method, the energy density goes down quickly as the gas is used up.

Figure 3 shows the cylinder and equivalent total gas volumes for a number of gases and for solid fuel. The highest equivalent total gas volumes are obtained with solid fuel combustion, followed by hydrogen and liquid ammonia. For safety reasons, however, nitrogen gas is often used.

Energy storage equipment using a flywheel has been developed experimentally and has achieved an energy density of 300 Wh/kg. However, this method has not yet reached a practical stage, and the gyroscopic effect and safety of the method are still in question.

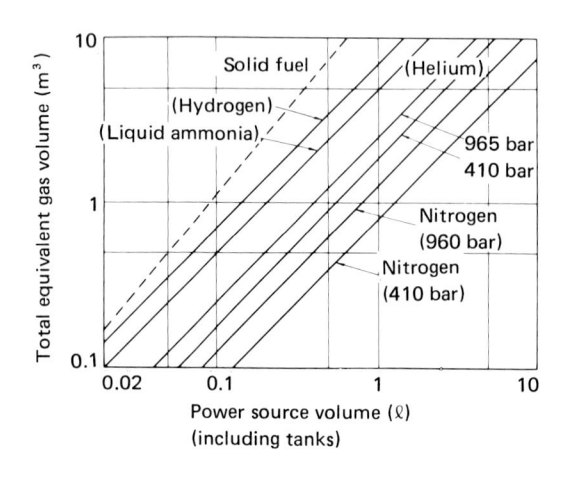

Figure 3 Gas bomb volume and equivalent storage gas volume.

Part 2

DRAWINGS

The following drawings are divided into human type and free type. The former type consists mainly of prosthetic hands, where the emphasis in design is put not only on function but also on shape. In the latter type, the functions are considered first and the design is less influenced by the shape of the human hand. Although many of the hands in these drawings may look similar, they are slightly different in accordance with their uses. These differences are noted below the drawings, in remarks related to the purpose of the design and the selling points.

HUMAN TYPE

CHUO HAND X1 | Chuo University

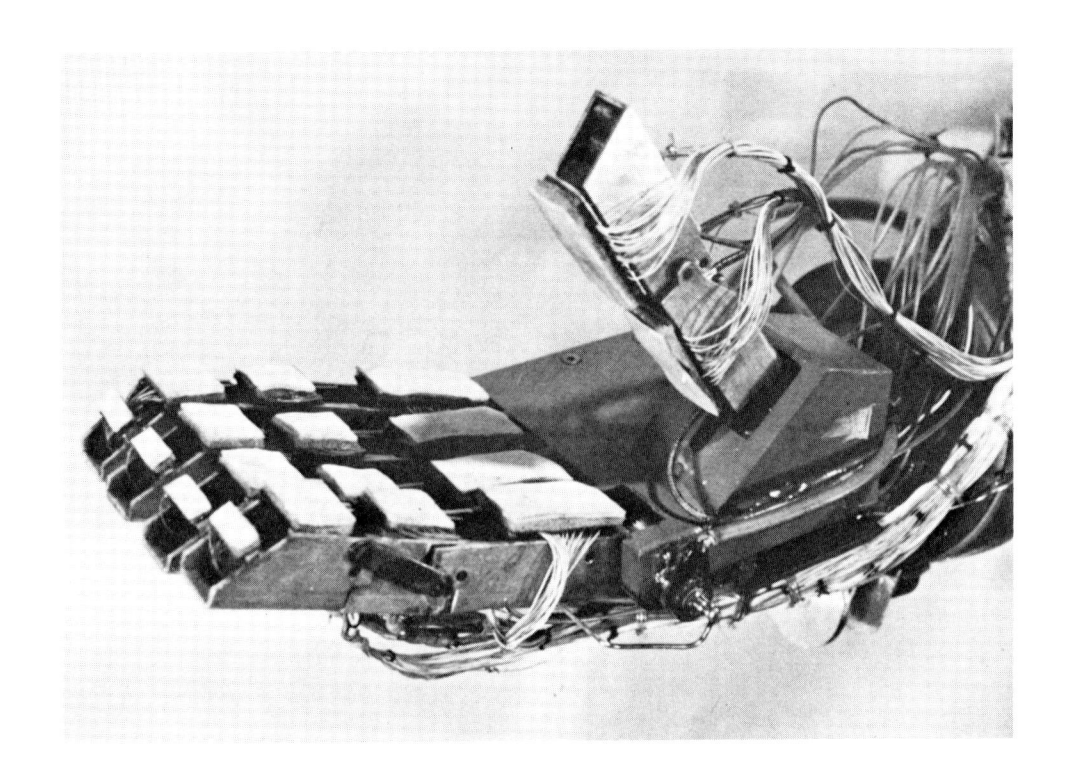

Working Speed				Working Range			
finger		wrist		finger		wrist	
open & shut		up & down	rotation	open & shut		up & down	rotation
free		freely settable		30°		20°	360°

Power Source	Sensors	Maximum Load	Size	Weight
Electricity (AC, DC servomotor), air pressure	Strain gauges, on-off switches		Total arm length 1180 mm	

NOTE

1. The hand was developed for research on recognizing the shape of an object by artificial touch sensors and investigating handling with five fingers. Using a computer, the forefinger metacarpophalangeal joints are controlled by an AC servomotor and the other joints are independently controlled by air cylinders. The artificial touch sensors consist of 384 switches.
2. The shoulder and elbow are controlled by on-off switches, using an AC servomotor and an air cylinder, respectively.
3. There are five fingers, and the base of the forefinger is controlled by an AC servomotor. The other joints are controlled by air cylinders.

FLEXIBLE MECHANICAL HAND

Electrotechnical Laboratory

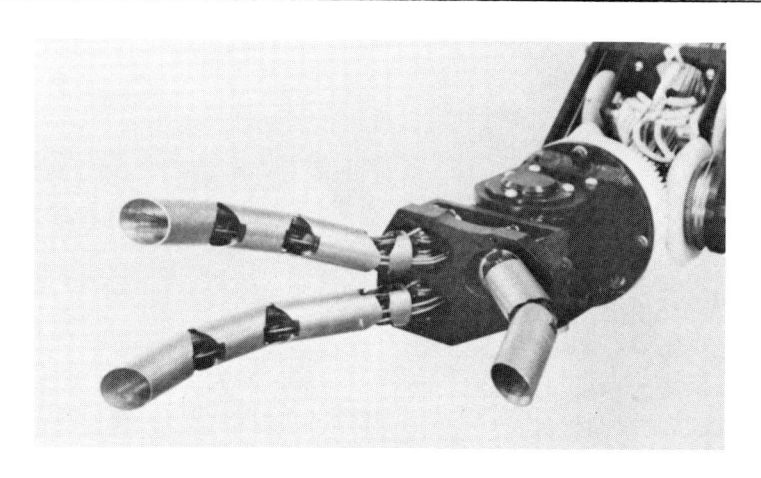

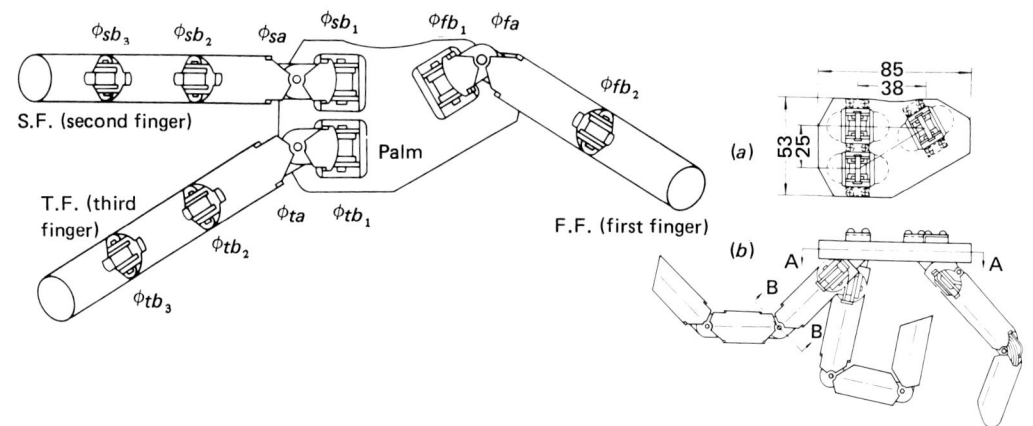

Working Speed			Working Range		
finger	wrist		finger	wrist	
open & shut	up & down	rotation	open & shut	up & down	rotation
500–600°/s	90°/s	45°/s	−45–90°	−60–60°	−180–180°

Power Source	Sensors	Maximum Load	Size	Weight
Electricity	None	500 g		240 g

NOTE

1. Compact multijoint hand characterized by high-level finger ability. Three fingers have stretch and side-bend joints. First finger has three degrees of freedom; second and third have four. The arm has five.
2. The moving ranges of joints are as follows. For the first finger, −45–45° at the first stretch joint, −45–90° at the second, −60–60° at the first side-bend joint. For the second finger, −45–45° at the first stretch joint, −45–90° at the second and third, −60–45° at the first side-bend joint. For the third finger, −45–45° at the first stretch joint, −45–90° at the second and third, −60–45° at the first side-bend joint.

HUMAN TYPE

WAM-4	Waseda University

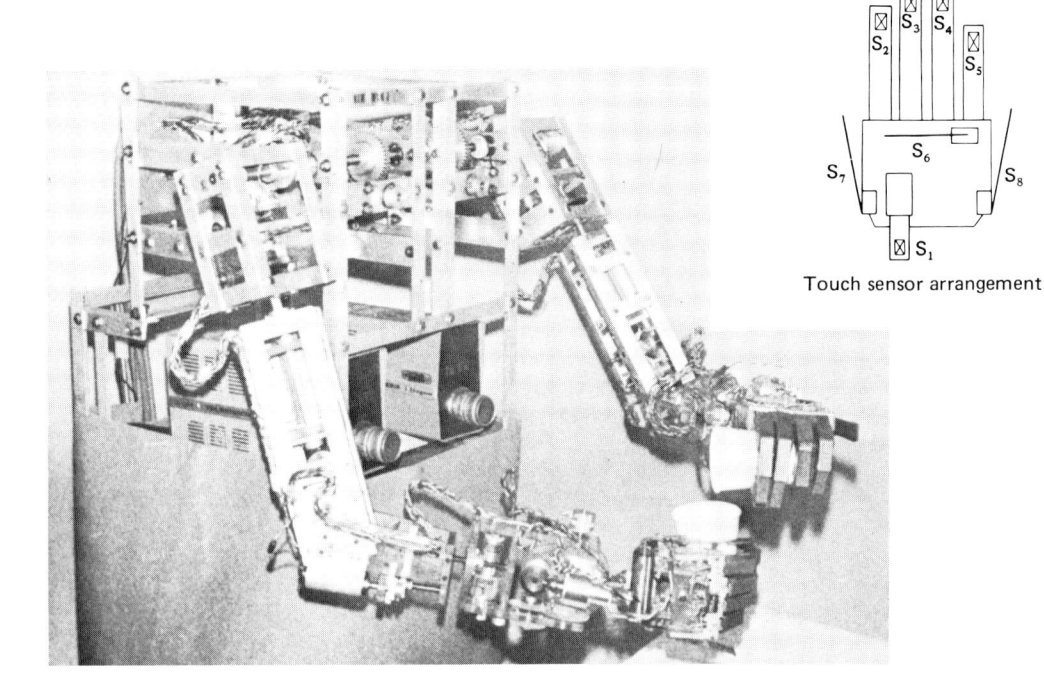

Touch sensor arrangement

Working Speed				Working Range			
finger	wrist		finger	wrist			
open & shut	up & down	rotation	open & shut	up & down	rotation	
1 s	8 rpm	8 rpm	90 mm	130 mm	130 mm

Power Source	Sensors	Maximum Load	Size	Weight
Stabilized DC	Vision (TV cameras), position (potentiometers), touch (8 micro switches for each arm)	500 g	Total length 765 mm, from shoulder to elbow 400 mm, from elbow to wrist 235 mm	20 kg

NOTE

1. Computer control lets sensors work together and perform operations according to a hierarchy program.
2. The hierarchy program was developed for searching bodies, exchanging the contents of paper cups, and so on. It consists of (1) a control program, (2) elemental working program modules, (3) elemental program modules, (4) typewriter control main program, and (5) typewriter control subprogram modules.
3. Manufactured for trial as the subsystem of a human-type robot.
4. Hands work cooperatively without redundancy. The relative positions of objects are recognized with the vision sensors, and the objects are searched, grasped, and manipulated with the touch sensors.

WAM-5 — Waseda University

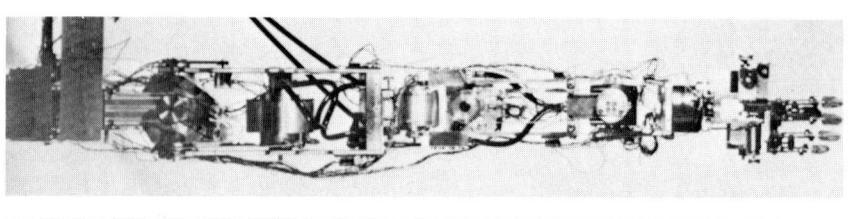

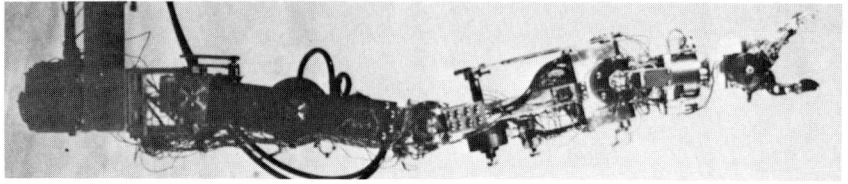

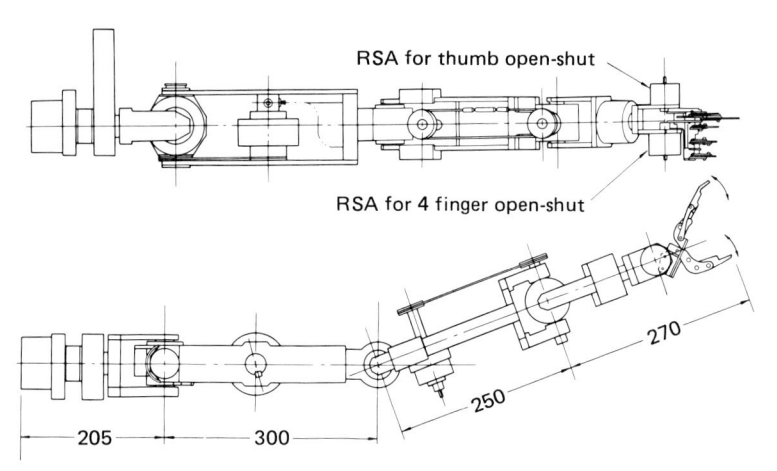

Working Speed			Working Range	
finger	wrist		finger	wrist
open & shut	up & down	rotation	bend	bend
			21 mm for thumb 87 mm for the other	−45–45°

Power Source	Sensors	Maximum Load	Size	Weight
Electricity	Position, force, touch, pressure	1 kg	1025 mm (total) (shoulder 205 mm, arm 300 mm, forearm 250 mm, hand 270 mm)	

NOTE

1. Human-type hand.
2. Three axes cross one axis at the shoulder and the wrist.
3. Degrees of rotation are arranged at both hands.
4. Potentiometers for position control and strain gauges for torque control are set at every joint.
5. One redundant degree of freedom for the position-attitude setting. Torque, touch, and pressure sensors are provided.
6. Introduces torque position control corresponding to external constraints.
7. Minicomputer control and torque position control.
8. −90–90° for shoulder horizontal rotation, −75–25° for shoulder inner and external rotation, −25–85° for shoulder bend, 20–120° for elbow bend.

HUMAN TYPE

WASEDA HAND 5 | Waseda University

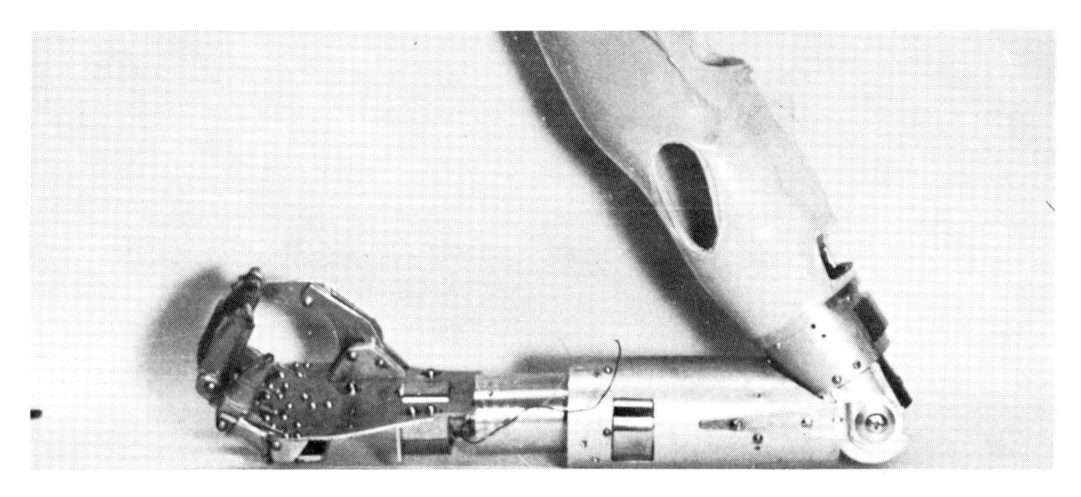

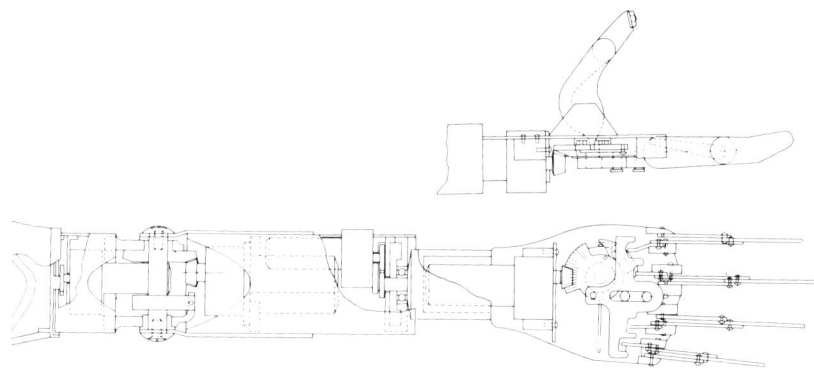

Working Speed			Working Range		
finger	wrist		finger	wrist	
open & shut	up & down	rotation	open & shut	up & down	rotation
10 s		4.5 rpm	90 mm		270°

Power Source	Sensors	Maximum Load	Size	Weight
Electricity	Force sense (Matsushita pressure sensitive diode at the thumb tip)	500 g	Total 470 mm (except socket), from elbow to hand 405 mm	2.1 kg

NOTE

1. Micro switches and electromyographic (EMG) combining control. Pressure feedback mechanism by electric stimulation.
2. Combination of paste electrodes, field effect transistor (FET) input stage, and class C amplifier increases both stability and signal-to-noise (S/N) ratio. Stimulation by pulsed current of MPS diode at the fingertip. Interference solved by the EMG selecting inputs in phase.
3. Artificial arm and hand having the same sizes and working ranges as human arm and hand.
4. Beautiful and easy to use because of the human-like five fingers. Grasp of a paper cup is stable because sensors control the grasping force and feedback to the living body.

ALL-ELECTRIC PROSTHETIC HAND (TD50-3)

University of Tokyo
Tokyo Denki University

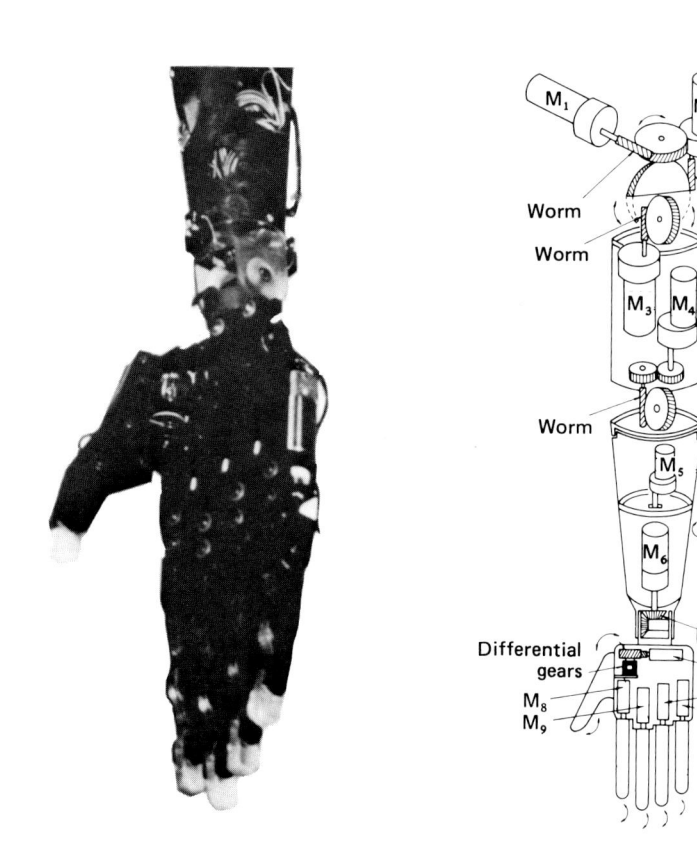

Working Speed			Working Range		
finger	wrist		finger	wrist	
open & shut	up & down	rotation	open & shut	rotation	bend
Same as the human finger	Same as human		Same as human	±90°	±90°

Power Source	Sensors	Maximum Load	Size	Weight
Electricity	None	About 1 kg		1.75 kg

NOTE

1. Voice control is achieved by a microcomputer. Cooperative motion of motors.
2. Pulse control of DC motors.
3. Made of carbon fiber, plastics, and light alloy.
4. Cooperative motion of motors is determined by computer processing of the numerical analysis of the arm's motion.
5. Modularization of shoulder, arm, elbow, wrist, and hand.
6. Working range of the shoulder is 180° for horizontal rotation, 130° for forward up-down, 90° for sideways up-down. That of the elbow is 120° for stretching, of the forearm is 270° for rotation, and of the wrist is ±90° for bend.

HUMAN TYPE

| TD-2 HAND | Tokyo Denki University |

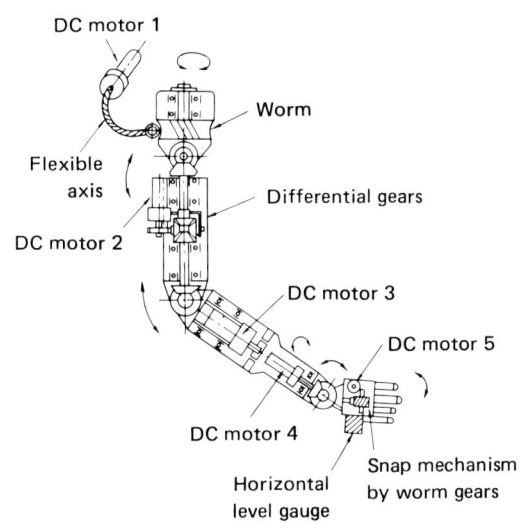

Labels in diagram: DC motor 1, Worm, Flexible axis, DC motor 2, Differential gears, DC motor 3, DC motor 5, DC motor 4, Horizontal level gauge, Snap mechanism by worm gears

Working Speed			Working Range		
finger	wrist		finger	wrist	
open & shut	up & down	rotation	open & shut	up & down	rotation
0.3 s/90°	0.2 s/90°	0.5 s/180°	0-90°	±90°	±180°

Power Source	Sensors	Maximum Load	Size	Weight
Electricity (DC motors)	Horizontal level holding of wrist level sensor	250 g	630 mm	2.1 kg

NOTE

1. Multirotation servosystem with potentiometers.
2. Orders transmitted with a joystick.
3. Ten-channel signals stored in magnetic tapes and played back.

ALL-ELECTRIC MECHANICAL HAND (TDH51-1)

University of Tokyo;
Tokyo Denki University

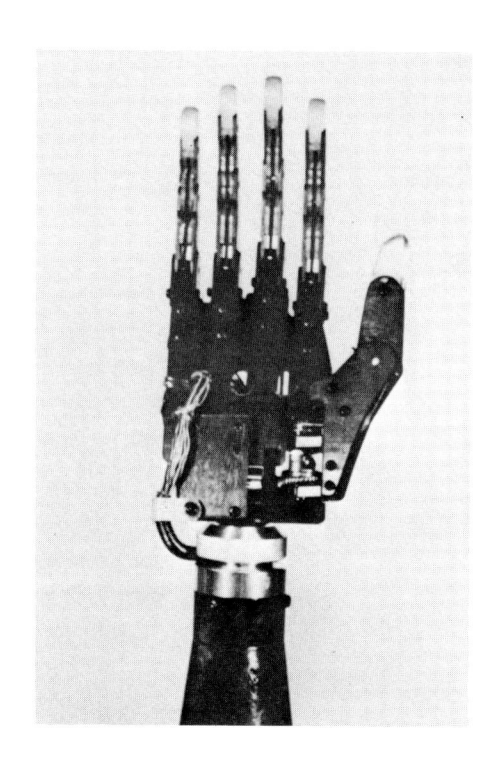

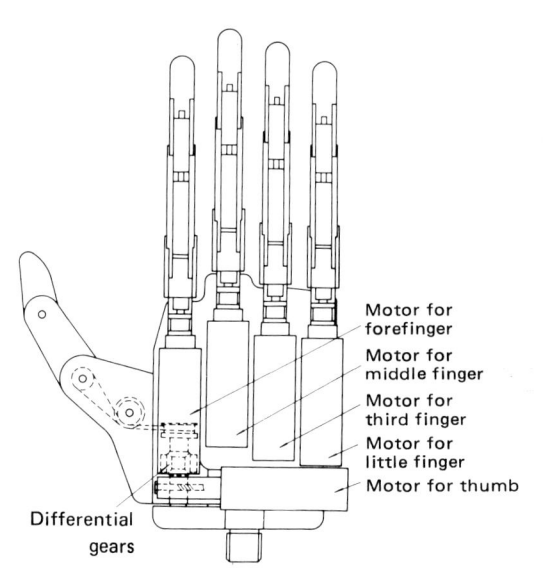

Motor for forefinger
Motor for middle finger
Motor for third finger
Motor for little finger
Motor for thumb
Differential gears

Working Speed			Working Range		
finger	wrist		finger	wrist	
open & shut	up & down	rotation	open & shut	up & down	rotation
Same as human fingers	Freely settable		180°	Freely settable	

Power Source	Sensors	Maximum Load	Size	Weight
Electricity	None	About 1 kg		0.5 kg

NOTE

1. Every finger has a motor.
2. Cooperative operation is possible.
3. Sensors are prepared to be attached.
4. Ranges for fingers are 90° for thumb internal and external rotation, 90° for thumb grasping, and 180° for other fingers grasping.

HUMAN TYPE

ELECTRIC PROSTHETIC FOREARM (TDU51-1,2)

University of Tokyo; Tokyo Denki University

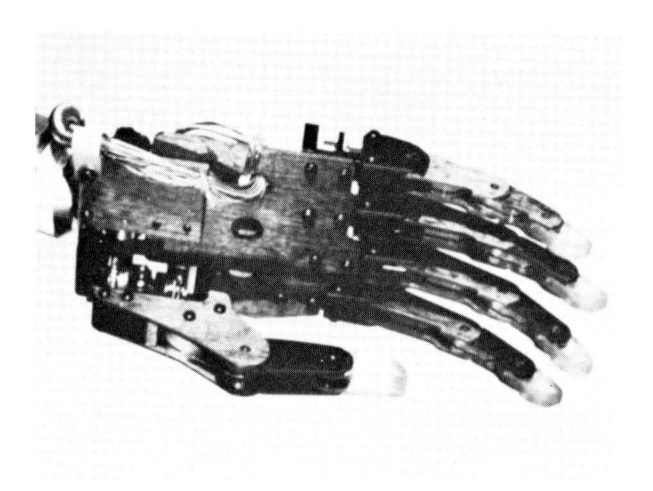

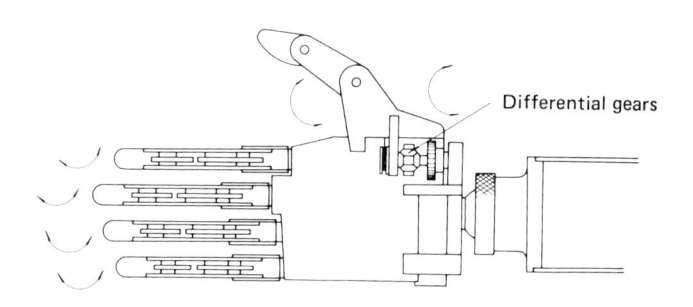

Differential gears

Working Speed				Working Range			
finger	wrist		finger	wrist		
open & shut	up & down	rotation	open & shut	up & down	rotation	
Same as human fingers			Same as human fingers	±90°	±90°	

Power Source	Sensors	Maximum Load	Size	Weight
Electricity	None	About 1 kg		1.5 kg

NOTE

1. Voice control achieved by a microcomputer. Cooperative motion of motors.
2. Pulse control of DC motors.
3. Made of carbon fiber, plastics, and light alloy.
4. Cooperative motion of motors determined by computer processing of the numerical analysis of the arm's motion.
5. Elbow is manually controlled.
6. Elbow, forearm, wrist, and fingers have degrees of freedom; 120° for elbow stretch and 120° for forearm internal and external rotation.

HUMAN-TYPE HAND	Mechanical Engineering Laboratory

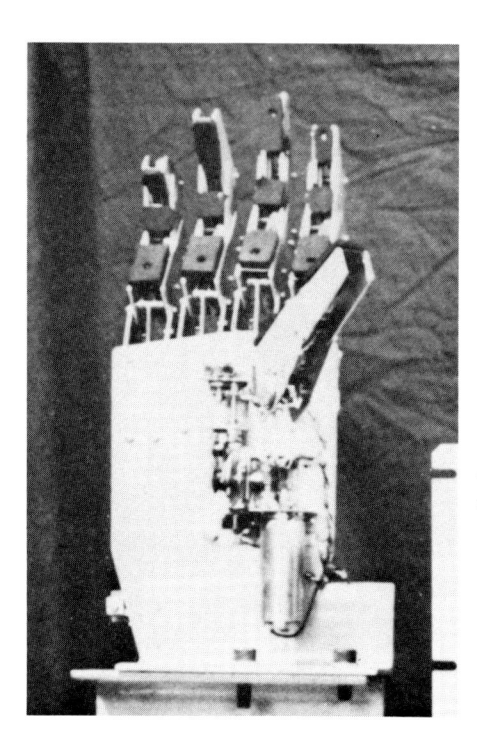

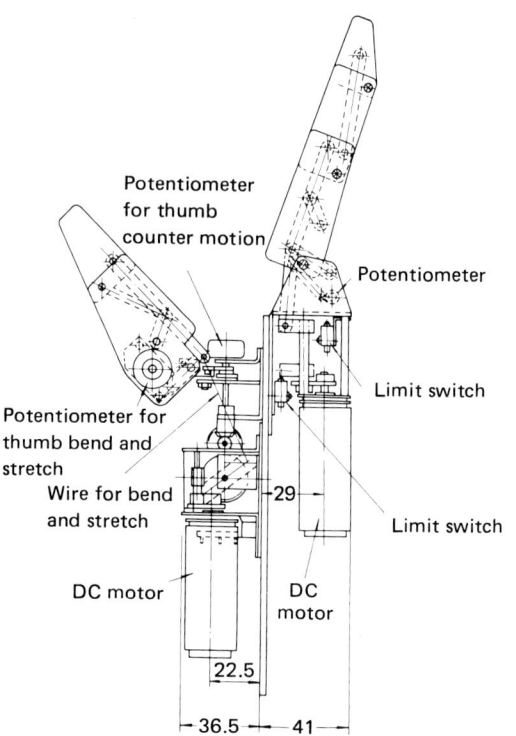

Potentiometer for thumb counter motion
Potentiometer
Potentiometer for thumb bend and stretch
Limit switch
Wire for bend and stretch
29
DC motor
Limit switch
DC motor
22.5
36.5 — 41

Working Speed			Working Range	
finger	wrist		finger	wrist
open & shut	up & down	rotation	open & shut	opposition
Freely settable	Freely settable		110 mm for thumb, 200 mm for other	100 mm for thumb

Power Source	Sensors	Maximum Load	Size	Weight
Electricity (DC motors)	Sense of touch (on-off switch for each finger section)	2.1 kg	115 mm for thumb, 115 mm for maximum of the others, 100 × 124 mm for back of hand	900 g

NOTE
1. Developed as a pair of hands for the human arm-type manipulator (MELARM).
2. Thumb has a motor and can make two motions.
3. Fingers have adaptive link mechanism and are driven by DC servomotors.
4. Can be operated by master-slave control, single-axis drive, or direct CPU control using 12-bit microprocessor.

HUMAN TYPE

WASEDA HAND 4 — Waseda University

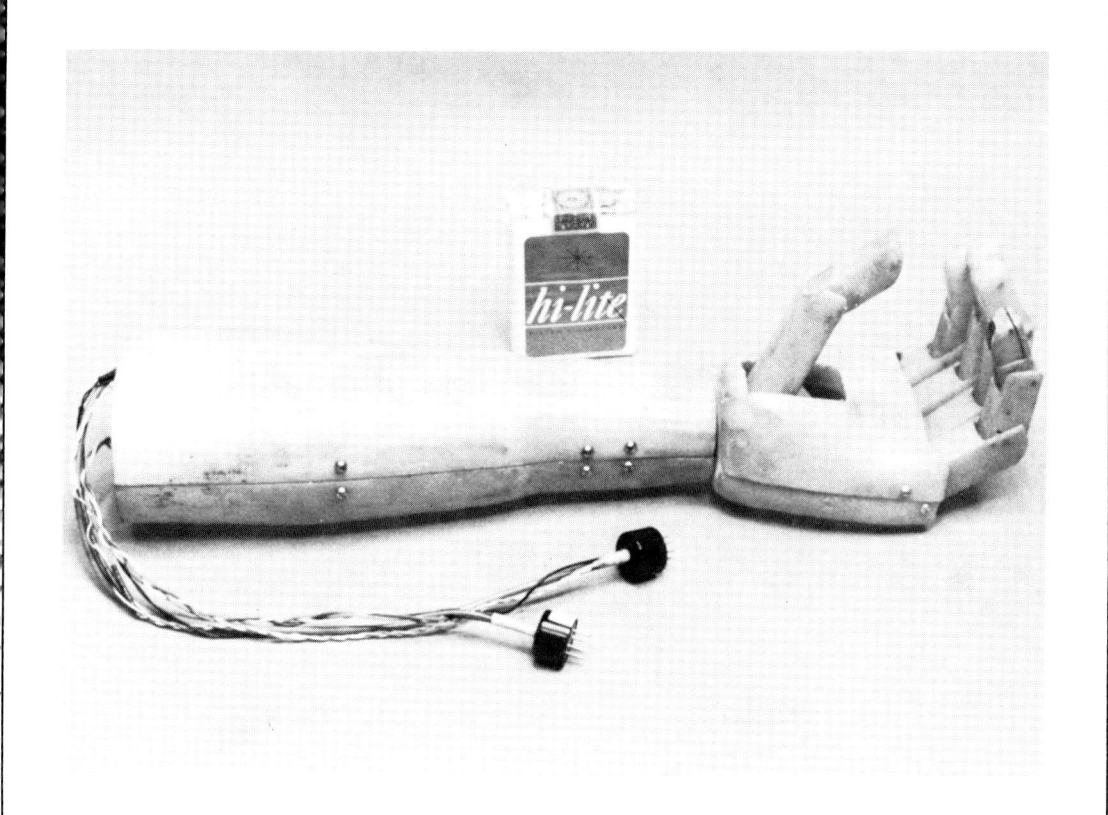

Working Speed			Working Range		
finger	wrist		finger	wrist	
open & shut	up & down	rotation	open & shut	up & down	rotation
1 s					

Power Source	Sensors	Maximum Load	Size	Weight
Battery	Pressure sensor (pressure element at the front of the forefinger)	5 kg	Same as human hand	880 g

NOTE

1. Pressure feedback control proportional to EMG.
2. Pressure feedback made by a mechanical vibrator stimulating the living body; the frequency is produced by converting the signal of the pressure sensor at the fingertip. The motor can generate torque proportional to the EMG by two-channel electrodes.

WASEDA IMASEN MYOELECTRIC HAND — Waseda University

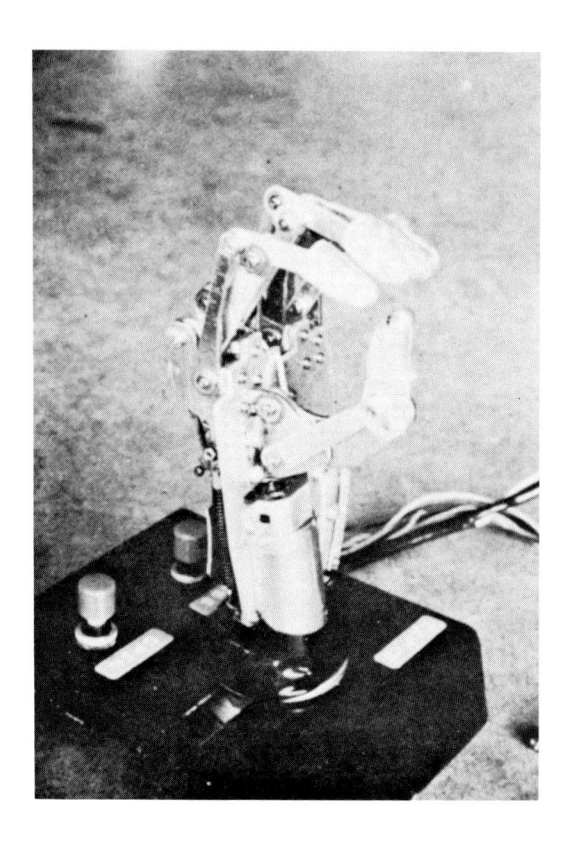

Working Speed			Working Range		
finger	wrist		finger	wrist	
open & shut	up & down	rotation	open & shut	up & down	rotation
1.1 s			110 mm		

Power Source	Sensors	Maximum Load	Size	Weight
Ni–Cd battery	Pressure sensors (strain gauges at the four fingers' driving screw spindles and bearings)	1.5 kg	Total length 198 mm, width 69 mm	940 g

NOTE

1. On–off control by EMG. Hand is opened and shut by the EMG at the remnant muscle (stretch and bend muscle). The EMG is rectified and smoothed. The gate opening by EMG order prevents errors.
2. Two types of grasping are possible: grasping with five fingers and picking up at three points, by mechanical flipflops.
3. Finger links and power convertor satisfy the specified strength, output, and appearance. Mechanical flipflops make grasping and three-point picking up possible.
4. Beautiful. Endurance several hundred thousand times the fatigue test value. Completed field test by 25 humans. Control circuit, battery power for a day, and optional force sensor.

HUMAN TYPE

KUMADAI HAND IV

Kumamoto University

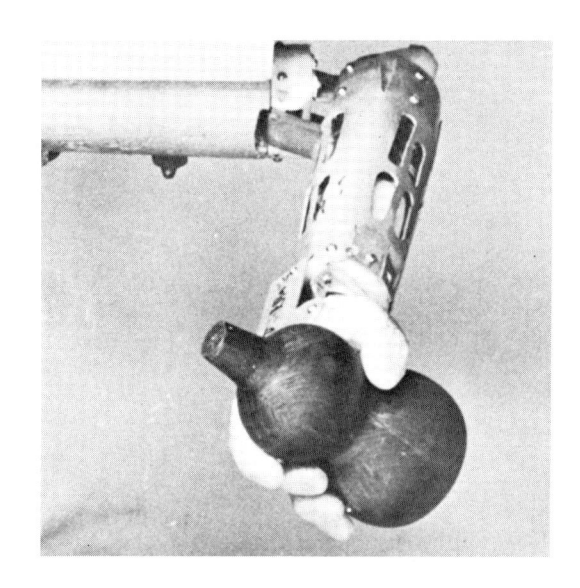

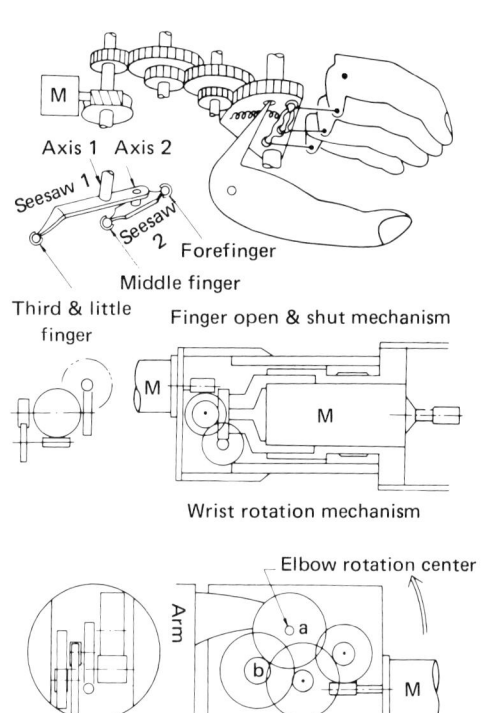

Axis 1 Axis 2
Seesaw 1
Seesaw 2
Forefinger
Middle finger
Third & little finger
Finger open & shut mechanism

Wrist rotation mechanism

Elbow rotation center
Arm
Elbow bend mechanism

Working Speed			Working Range		
finger	wrist		finger	wrist	
open & shut	up & down	rotation	open & shut	up & down	rotation
75°/2 s					

Power Source	Sensors	Maximum Load	Size	Weight
Electricity (battery)	None	1.2 kg; grasp force is 4 kg		720 g (contained motor weight, 250 g)

NOTE

1. Controlled by switches.
2. Total efficiency 15–20%, mechanical efficiency 30–40%.
3. Made of brass, aluminum, and plastics for non-oil bearings.

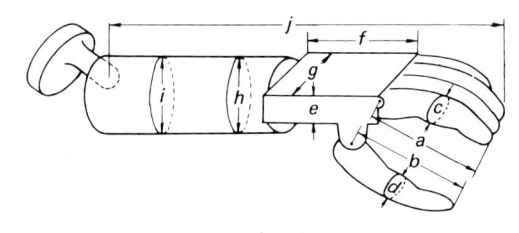

a	b	c	d	e	f	g	h	i	j
73	81	12	15	25	72	54	45	58	310

NEW 10H7 WITH 7 DEGREES OF FREEDOM POWERED BY OIL PRESSURE

Waseda University

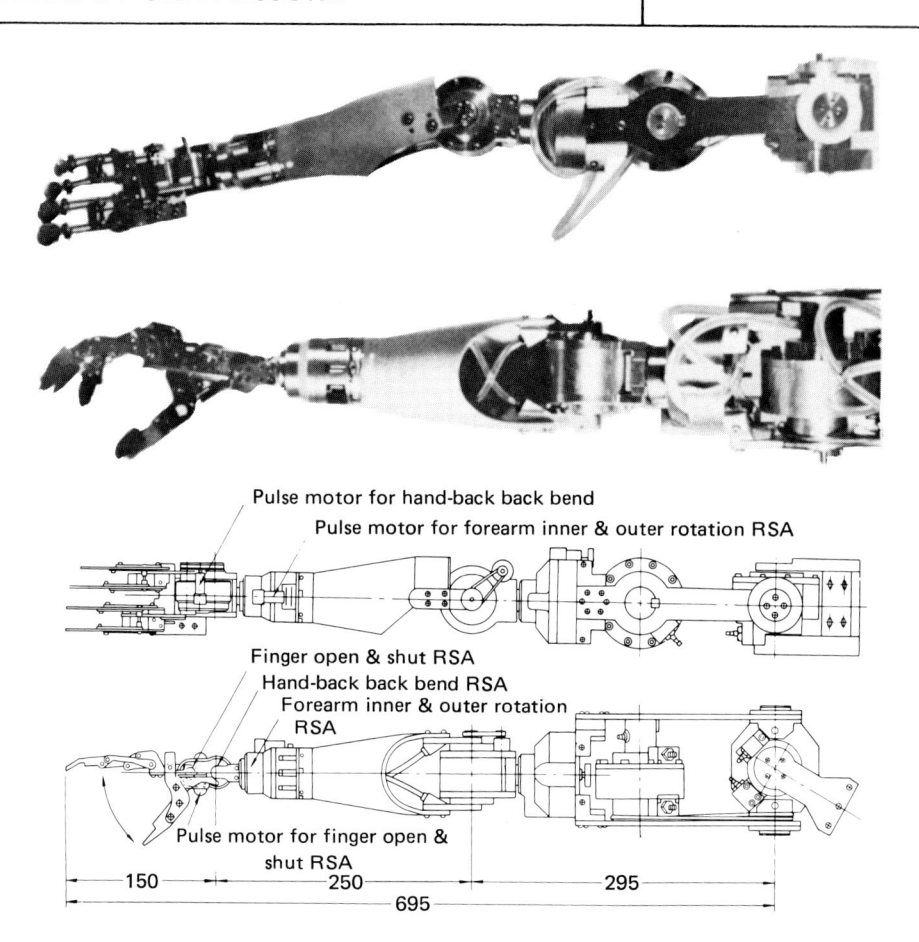

Pulse motor for hand-back back bend
Pulse motor for forearm inner & outer rotation RSA

Finger open & shut RSA
Hand-back back bend RSA
Forearm inner & outer rotation RSA

Pulse motor for finger open & shut RSA

150 — 250 — 295
695

Working Speed			Working Range		
finger	wrist		finger	wrist	
open & shut	up & down	rotation	open & shut	up & down	rotation
1.0 s	90°/s	90°/s	open width 110 mm	90° (from −45° to 45°)	90° (from −30° to 60°)
Power Source	**Sensors**	**Maximum Load**	**Size**	**Weight**	
Oil pressure	None	1 kg	Total length 695 mm	3.4 kg	

NOTE

1. Increased power of the wrist. Both degrees of freedom and internal oil pipe system improved.
2. Hardware made for usual operation and its evaluation.
3. Minicomputer control (microcomputer control in future).
4. EMG patterns at the shoulder specify the hand's motion.
5. Cooperative control adopted. Actuator does not go out from the arm. Total oil pressure system.

HUMAN TYPE

| PROSTHETIC HAND WITH 7 DEGREES OF FREEDOM AT THE SHOULDER POWERED BY OIL PRESSURE | Waseda University |

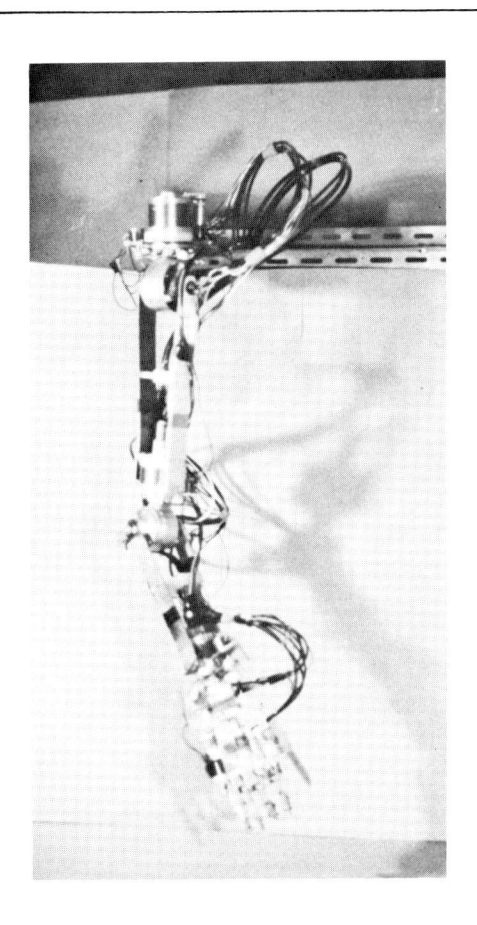

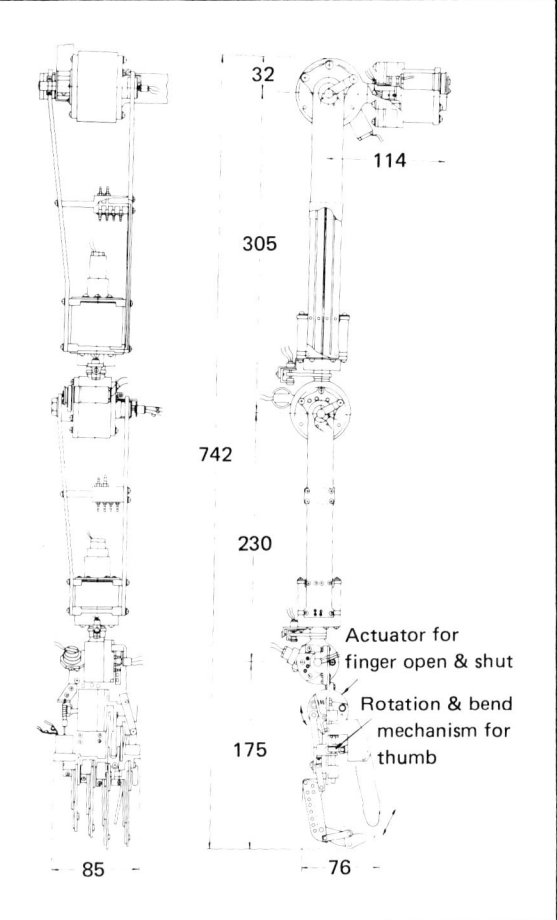

Actuator for finger open & shut
Rotation & bend mechanism for thumb

Working Speed				Working Range			
finger	wrist		finger	wrist			
open & shut	up & down	rotation	open & shut	up & down	rotation
0.5 s	33 rpm	35 rpm	66 mm	90°	300°

Power Source	Sensors	Maximum Load	Size	Weight
Oil pressure	Position potentiometers (each degree of freedom has one from shoulder to wrist; total of six)	1 kg	Total length 645 mm, from shoulder to elbow 290 mm, from elbow to wrist 230 mm	2.5 kg

NOTE

1. Microcomputer control (under development, now minicomputer control).
2. Eleven kinds of EMG pattern on the shoulder are made to correspond to the control modes and are selected by the human.
3. Control modes consist of cooperative modes of the hand and wrist. Cooperative control is terminal control.
4. Small, light RSA. Internal oil pipe system under investigation.
5. Lightness of the hand is due to the RSA.

PROTOTYPE OF PROSTHETIC UPPER ARM POWERED BY OIL PRESSURE

Waseda University

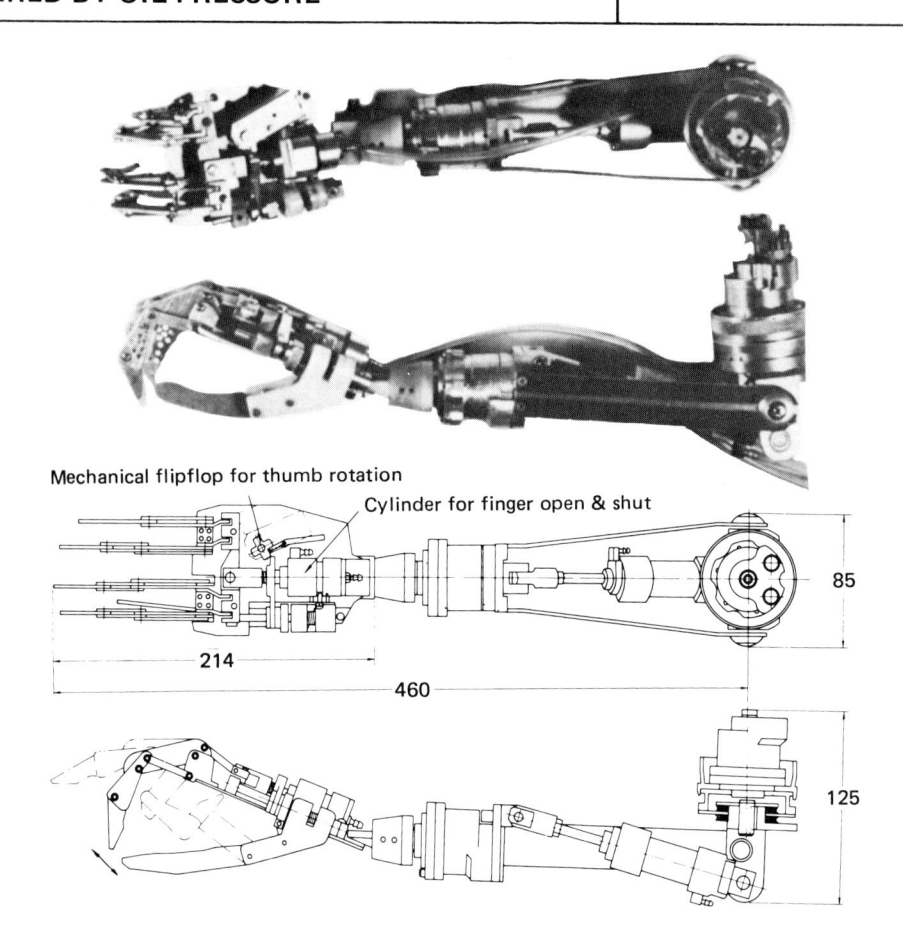

Working Speed			Working Range		
finger	wrist		finger	wrist	
open & shut	up & down	rotation	open & shut	up & down	rotation
1 s		192 rpm	80 mm (max.)		180°

Power Source	Sensors	Maximum Load	Size	Weight
Oil pressure (super-miniature pump)	None	3–4 kg	Total length 460 mm, width 85 mm, fingertip to wrist 214 mm	1400 g

NOTE

1. Independent velocity control for each degree of freedom.
2. Double oil actuator switched by very small direction valve, 4 ports—3 positions closed center type, and velocity control by pulse width modulator (PWM) is possible.
3. Hand lightened by using oil pressure. PWM control proportional to oil pressure being investigated. Plans to use carbon fiber beams.
4. Side grasping increases the function of the hand by a mechanical flip mechanism for thumb rotation.

HUMAN TYPE

PROSTHETIC FOREARM WITH 3 DEGREES OF FREEDOM POWERED BY BOTH ELECTRICITY AND OIL PRESSURE

Waseda University

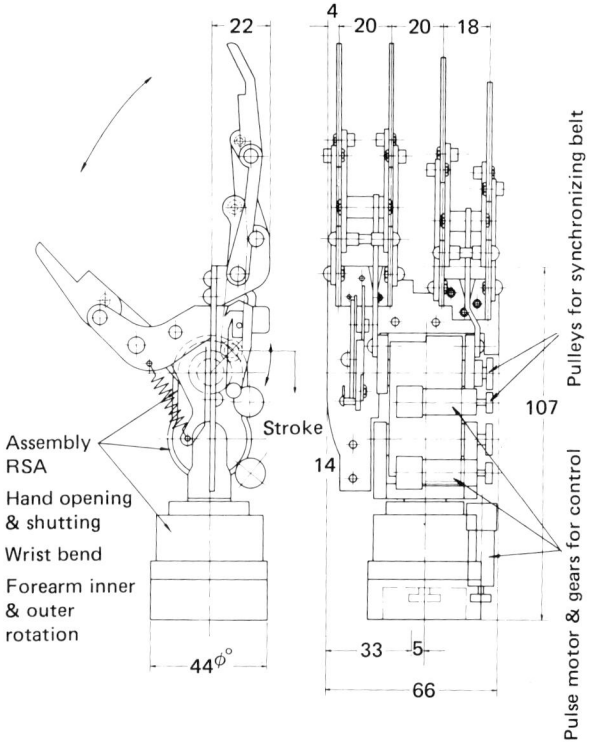

Assembly RSA
Hand opening & shutting
Wrist bend
Forearm inner & outer rotation

Stroke
Pulleys for synchronizing belt
Pulse motor & gears for control

Working Speed				Working Range			
finger	wrist		finger	wrist	
open & shut	up & down	rotation	open & shut	up & down	rotation
1 s	90°/s	90°/s	90°	90°	90°

Power Source	Sensors	Maximum Load	Size	Weight
Oil pressure	None	1 kg	Total length 216 mm, total width 66 mm, fingertip to wrist 148 mm	700 g

NOTE

1. Hand small and light due to both the compound RSA and the internal oil pipe system.
2. More degrees of freedom and lighter than usual hands. Control using EMG pattern eliminates special training.
3. Power source is a portable, extremely small oil pressure unit.
4. Signal for control is EMG at the remnant muscle. Both independent control of each degree of freedom and cooperative control between the wrist bend and two degrees of freedom of forearm rotation are exerted by using a recognition circuit and logic circuit. Each actuator is controlled with the open loop made by a pulse motor and an oil pressure torque amplifier.

| PROSTHETIC HAND WITH 7 DEGREES OF FREEDOM POWERED BY OIL PRESSURE (PROTOTYPE I) | Mechanical Engineering Laboratory |

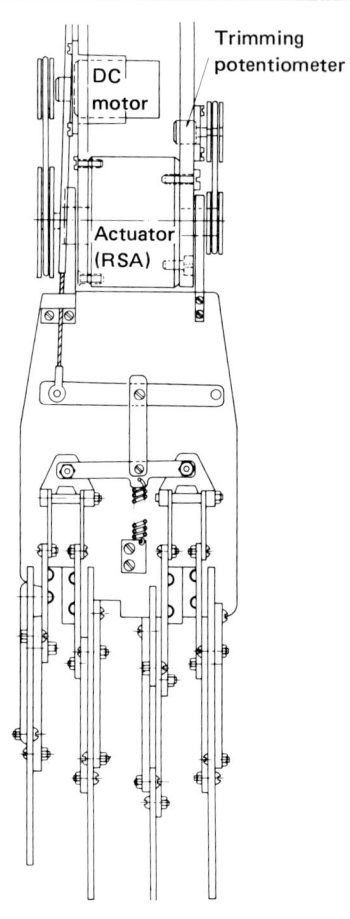

Working Speed			Working Range		
finger	wrist		finger	wrist	
open & shut	up & down	rotation	open & shut	back bend	forearm rotation
Freely settable	Freely settable		Free	180°	280°

Power Source	Sensors	Maximum Load	Size	Weight
Oil pressure	None	2 kg	670 mm	4 kg

NOTE

1. Assuming control by EMG, now developing software and on–off switching.
2. Planning to attach pressure and touch sensors.

INSTRUMENT FOR FINGER POWERED BY GAS PRESSURE

Tokushima University

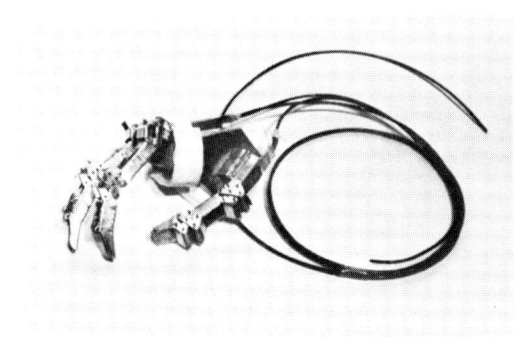

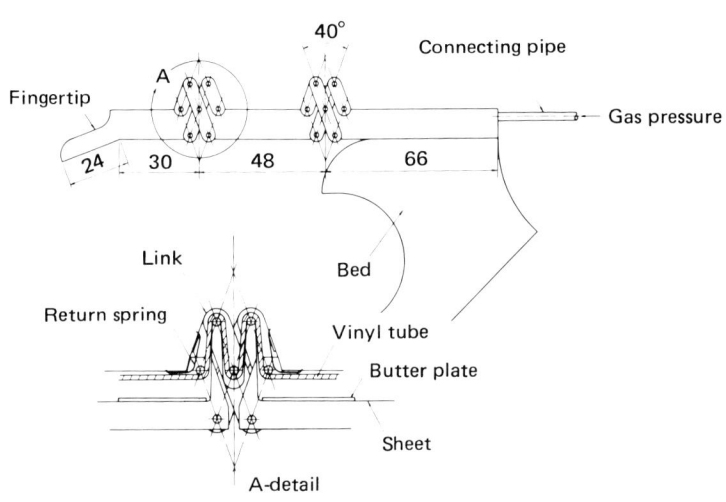

Working Speed			Working Range		
finger	wrist		finger	wrist	
open & shut	up & down	rotation	open & shut	up & down	rotation
2 s			70 mm	Free	

Power Source	Sensors	Maximum Load	Size	Weight
Gas pressure (acid gas), compressed air	None	Grasping force 500 g (at 2 kg/cm^2 gas pressure)	Changeable according to size of hand	700 g (containing an instrument, a small manual valve, etc.)

NOTE

1. These instruments are now outmoded. An instrument for humans with paralyzed fingers must be light. This is attached to the back of the hand and opened and shut by use of acid gas.
2. The light actuator made of vinyl tubing makes the instrument portable. This link mechanism can work smoothly without displacement between the finger and the tool.
3. Finger bending is achieved by compressing gas from a gas bomb into the vinyl tubing through a reducing and a driving valve (manual or electromagnetic).
4. Finger stretching is achieved by return springs and releasing gas in the tube.

| HAND WITH THREE FINGERS | Kyoto University |

Working Speed				Working Range	
finger	wrist				
open & shut	up & down	rotation	angle		distance
0.5 s			90°		75 mm
Power Source	Sensors	Maximum Load	Size	Weight
Electricity (motors)	Grasp force sensor and finger opening angle sensor. Three small potentiometers for each sensor.	4 kg		

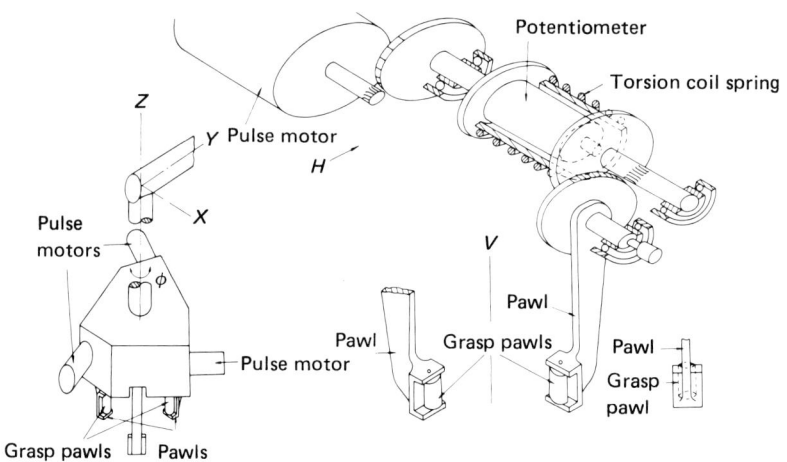

NOTE

1. Three fingers open symmetrically. They are driven by pulse motors through torsional coil springs, which produce pawl grasping force proportional to the torsional angle. Grasping force is controlled by using the potentiometers measuring the angle.
2. Cooperation of the three fingers to achieve complex grasp with stability and arbitrary rigidity.
3. Three electric pulse motors used as actuators.
4. Maximum grasping force 4 kg.

FREE TYPE: PICKING

EXTERNAL GRASPING SWING LEVER TYPE HAND

Yasukawa Electric Mfg. Co., Ltd.

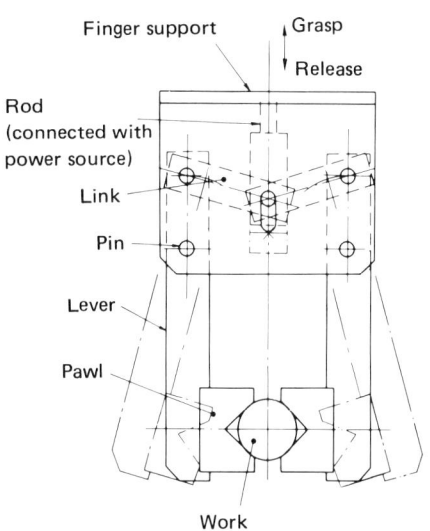

Finger support — Grasp / Release
Rod (connected with power source)
Link
Pin
Lever
Pawl
Work

Working Speed			Working Range		
finger	wrist		finger	wrist	
open & shut	up & down	rotation	open & shut	up & down	rotation
1 s (1.5 s by standard timer)	Freely settable		About 120 mm	Freely settable	

Power Source	Sensors	Maximum Load	Size	Weight
Electricity (motors)	None	About 30 kg	120 mmϕ (work)	About 6 kg

NOTE

1. Compact.
2. Maximum force can be produced, whenever necessary, by a toggle mechanism.
3. By exchanging pawls several kinds of work can be grasped.
4. On–off control by timers and limit switch.

TDG–1 HAND | Tokyo Denki University

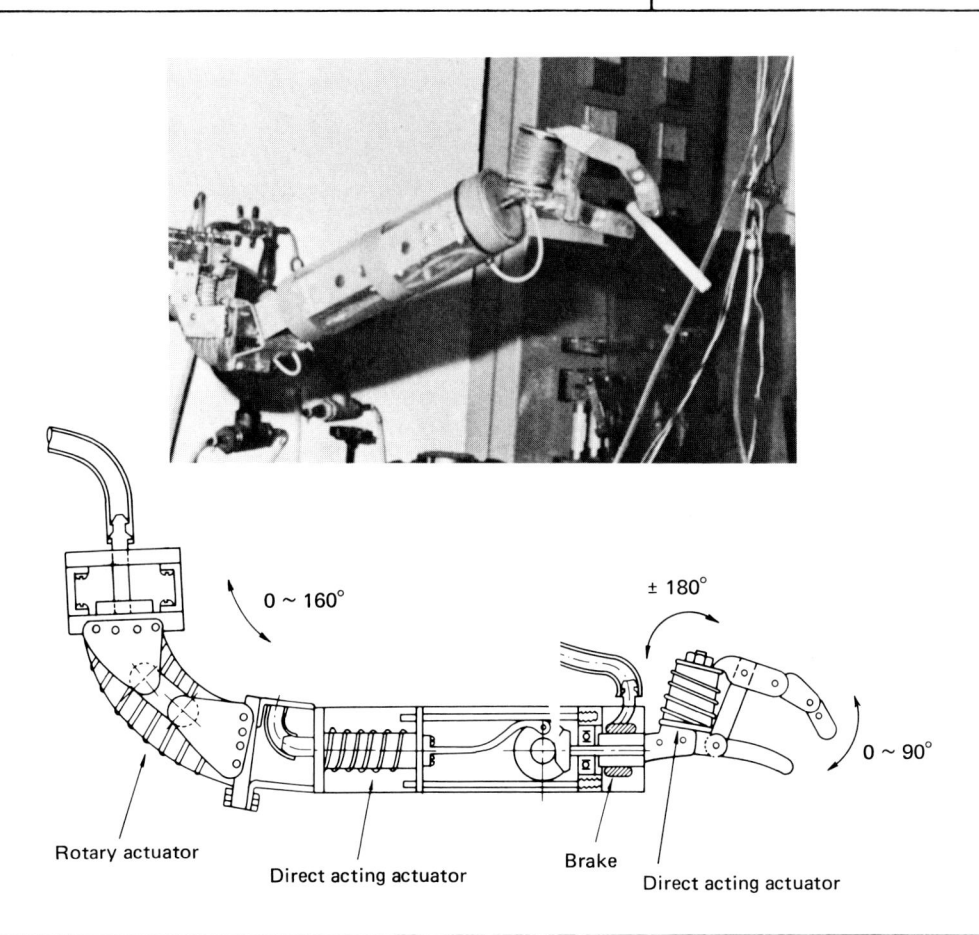

Rotary actuator
Direct acting actuator
Brake
Direct acting actuator

0 ~ 160°
± 180°
0 ~ 90°

Working Speed			Working Range		
finger	wrist		finger	wrist	
open & shut	up & down	rotation	open & shut	up & down	rotation
0.2 s/90°	0.3 s/180°		90° (max.)	±180°	

Power Source	Sensors	Maximum Load	Size	Weight
Compressed air (5 kg/cm² G)	None	300 g	550 mm	830 g

NOTE
1. On–off control.
2. Switching by electromagnetic valve and control by solenoids and air brakes.
3. Easy rotation by a rotary actuator.
4. The hand's working range is 160° right-left, and the arm's up-down working speed is 0.4 s.

FREE TYPE: PICKING 59

HAND FOR UPSETTING

Kawasaki Heavy Industries, Ltd.

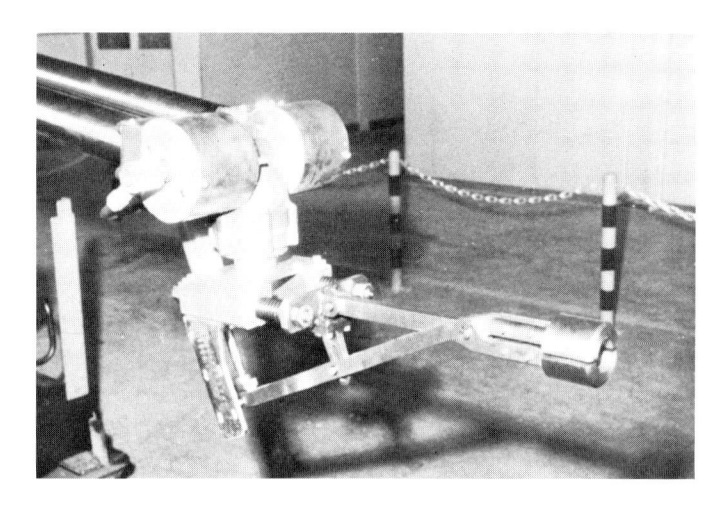

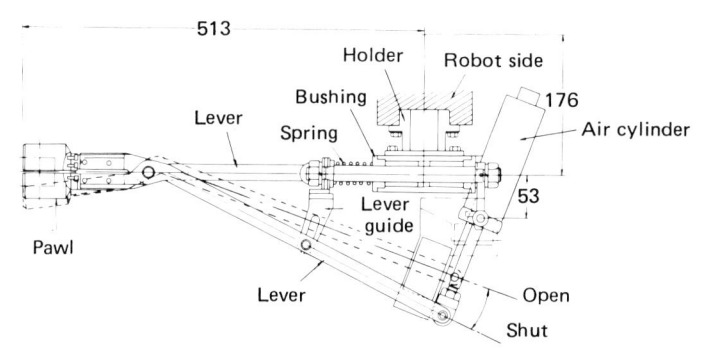

[Photograph of the hand for upsetting mechanism]

[Diagram showing components: 513 mm overall, Holder, Robot side, Bushing, Lever, Spring, 176, Air cylinder, Lever guide, 53, Pawl, Lever, Open, Shut]

Working Speed			Working Range		
finger	wrist		finger	wrist	
open & shut	up & down	rotation	open & shut	up & down	rotation
	Freely settable		Dependent on works	Freely designable	

Power Source	Sensors	Maximum Load	Size	Weight
Air pressure (5 kg/cm² G)	None	5–10 kg	513 mm in arm length	6 kg

NOTE

1. The upsetting with holding a work makes the working cycle time short, so the hand has a shock absorber mechanism for the machine side shocking force.
2. The upsetting with holding a work gives the pawl a thrust shock force, which is absorbed by springs.
3. On-off control by air pressure.

REMOVABLE HAND	Kawasaki Heavy Industries, Ltd.

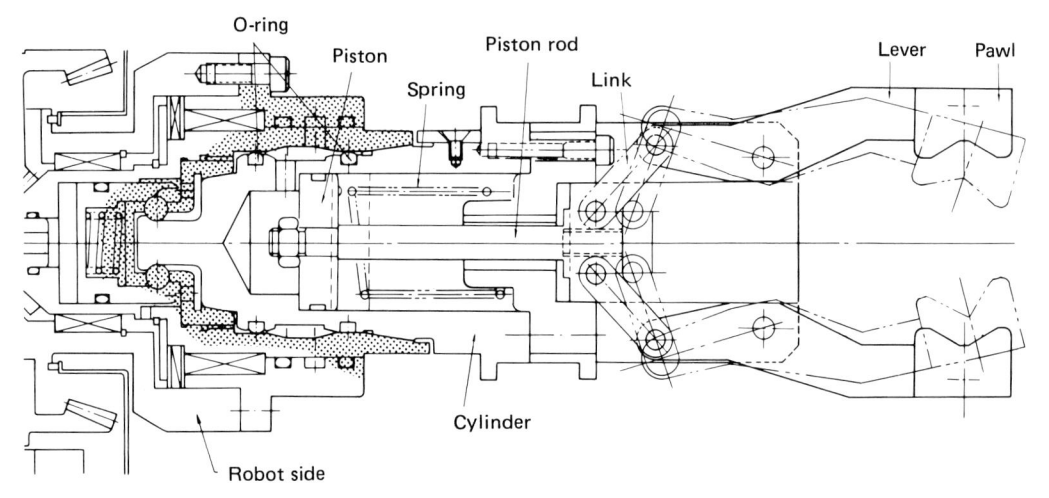

Labels: O-ring, Piston, Piston rod, Spring, Link, Lever, Pawl, Cylinder, Robot side

Working Speed			Working Range		
finger	wrist		finger	wrist	
open & shut	up & down	rotation	open & shut	up & down	rotation
	Freely settable		Dependent on work	Freely designable	

Power Source	Sensors	Maximum Load	Size	Weight
Air pressure (5 kg/cm^2 G)	None	1–5 kg	Dependent on work	4 kg

NOTE

1. The robot has the capability for self-exchange of the hand as shown above. The air circuit is automatically connected in the exchange.
2. Both the exchange and the chuck are checked electrically (utility model patent applied for in Japan).
3. On–off control by air pressure.

FREE TYPE: PICKING

| DOUBLE HAND (FOR MACHINE TOOLS) | Kawasaki Heavy Industries, Ltd. |

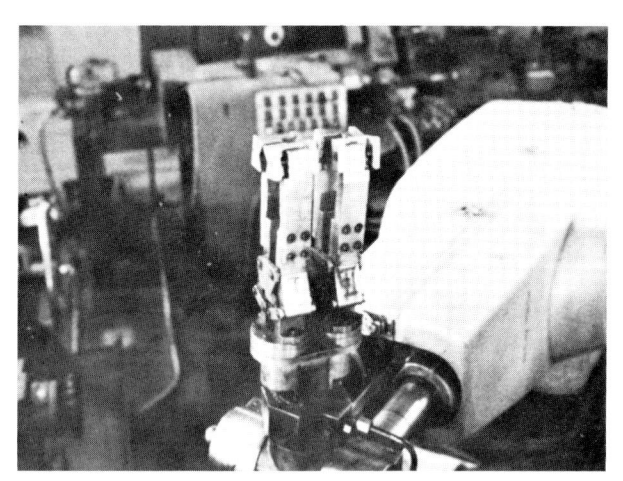

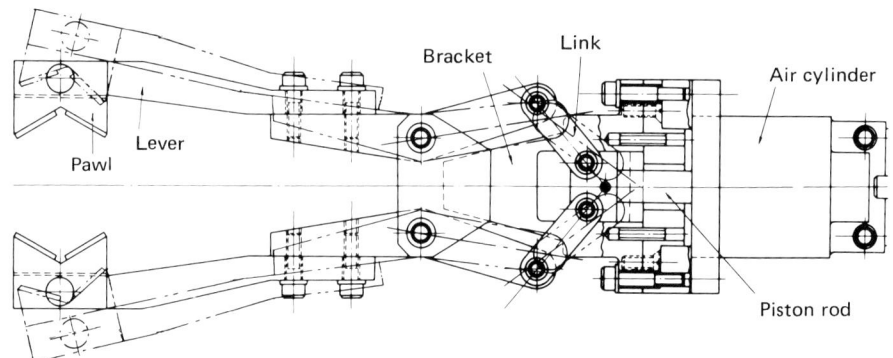

Working Speed				Working Range			
finger		wrist			finger		wrist	
open & shut		up & down	rotation		open & shut		up & down	rotation
Free		Freely settable			Dependent on work		Freely designable	

Power Source	Sensors	Maximum Load	Size	Weight
Air pressure (5 kg/cm² G)	None	2–3 kg	Dependent on work	About 8 kg

NOTE

1. Use of a double hand for loading and unloading shortens the idle time of a machine tool.
2. Each hand runs independently.
3. The robot has an air circuit inside, which eliminates pipes in the hand.
4. On–off control by air pressure.
5. One hand loads new work just after the other unloads preceding work from the machining center.

HAND FOR FORGING WORK

Kawasaki Heavy Industries, Ltd.

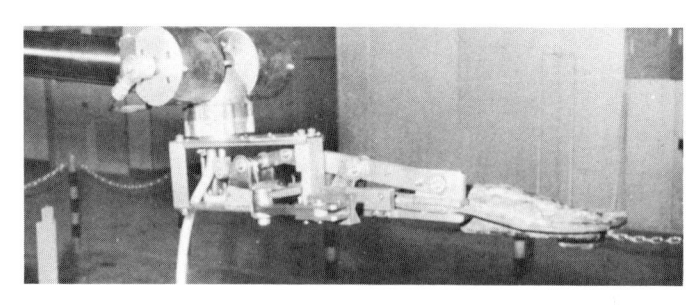

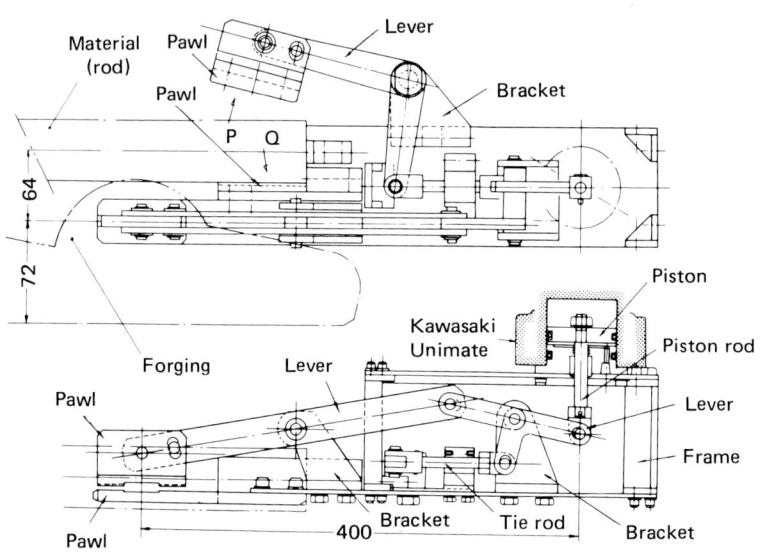

Working Speed				Working Range			
finger	wrist			finger	wrist		
open & shut	up & down		rotation	open & shut	up & down		rotation
	Freely settable				Freely designable		

Power Source	Sensors	Maximum Load	Size	Weight
Air pressure (5 kg/cm² G)	None	Under 20 kg	430 × 64 mm	12 kg

NOTE

1. The hand can handle deforming work from a material through three processes (crushing, forming, and trimming).
2. Two parts of the clamp run by one cylinder. The clamp fitting for the transformation is used in each process.
3. Compact for forging die.
4. One chuck for the material (a rod) and the other for the work flash (Japanese patent 1090620).
5. On-off control by air pressure.

FREE TYPE: PICKING 63

GRIPPER FOR TWO ARTICLES AT A TIME | Showa Kuatsuki Co., Ltd.

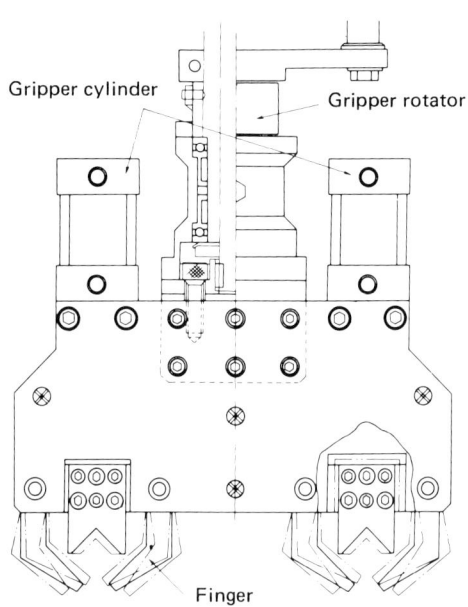

Working Speed				Working Range			
finger		wrist		finger		wrist	
open & shut		up & down	rotation	open & shut		up & down	rotation
30°/s		120°/s	15°			180°	

Power Source	Sensors	Maximum Load	Size	Weight
Air pressure	None	4 kg (1 kg for each)	Gripping part 20 or 30 mmϕ	400 kg (main robot)

NOTE

1. This is used as the loading robot in assembly; the wrist can rotate 180° to reverse the assembled work. Two sizes for the gripping unit diameter produce concentricity of two kinds of work by exchanging the gripping unit.
2. Ordering: sequence control by a pinboard matrix.
3. Positioning: mechanical stopping method.
4. Time: timer.
5. Velocity: preset method.

GRIPPER FOR GEAR CONVEYING

Showa Kuatsuki Co., Ltd.

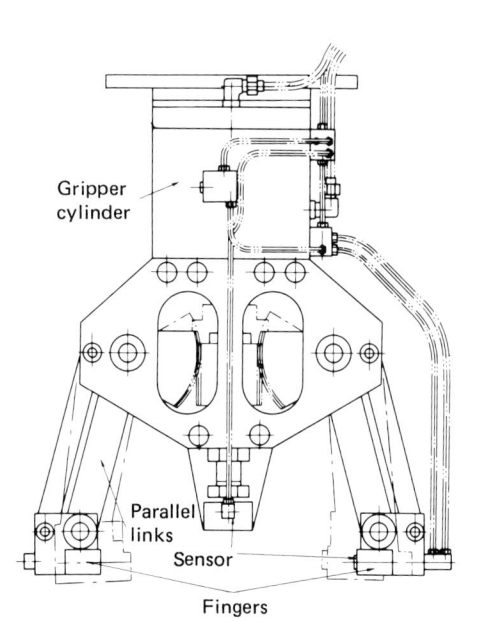

Working Speed			Working Range		
finger	wrist		finger	wrist	
open & shut	up & down	rotation	open & shut	up & down	rotation
20°/s			20°		

Power Source	Sensors	Maximum Load	Size	Weight
Air pressure	Air sensors	1 kg	150 mm O/ × 15 mm (thickness)	1500 kg (main robot)

NOTE

1. The gripper grasps a disk about 15 mm thick, adopting parallel links for various diameters of the work.
2. Rack and pinion gear makes centering possible; air sensor used to determine whether it clamps.
3. Ordering: semilocked sequence control.
4. Positioning: mechanical stopping method.
5. Time: timer.
6. Velocity: position method.

FREE TYPE: PICKING

HAND FOR DRILL LOADING

Shinko Electric Co., Ltd.

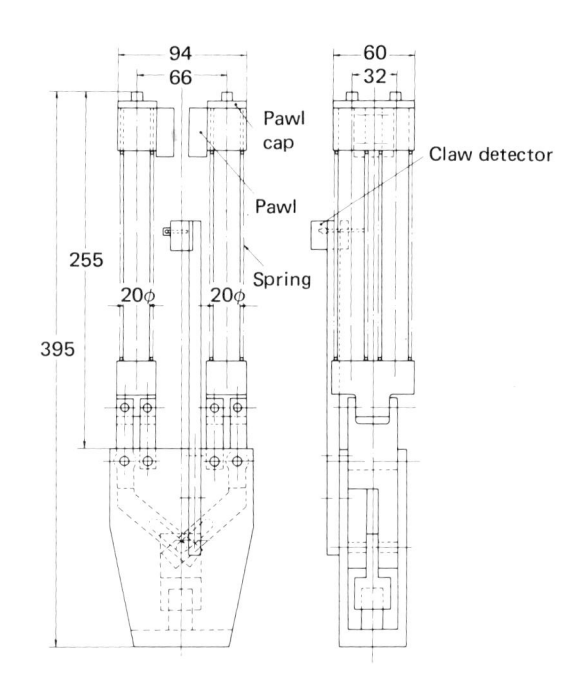

Working Speed			Working Range		
finger	wrist		finger	wrist	
open & shut	up & down	rotation	open & shut	up & down	rotation
0.4 s	Freely designable		Free	Freely designable	
Power Source	**Sensors**		**Maximum Load**	**Size**	**Weight**
Air pressure	None		Freely designable	395 (L) × 94 (W) × 60 mm (T)	

NOTE

1. Loading a drill for drill grinding machine.
2. Pawl escape mechanism used to hold drill rigidly until it is set in the chuck.
3. Parallel link jaw driven by one air cylinder.
4. The arm working range is 100 mm up and down, 500 mm back and forth, and 90° right and left; speed is 300 mm/s up and down, 500 mm/s back and forth, and 90°/s right and left.

PICK AND PLACE UNIT 1 — Nakamura-kiki Engineering

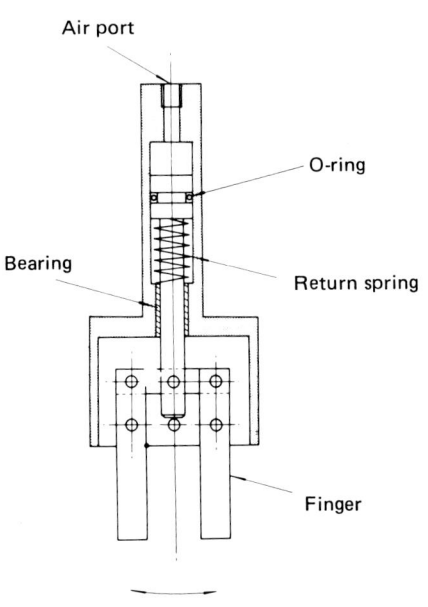

Working Speed			Working Range		
finger	wrist		finger	wrist	
open & shut	up & down	rotation	open & shut	up & down	rotation
0.3 s	Free		15°	Freely designable	

Power Source	Sensors	Maximum Load	Size	Weight
Air pressure	None	200 g (max.)	170 (L) × 82 (W) × 200 (H) mm	5.1 kg

NOTE

1. Toggle link method makes the gripping force strong in spite of the small size. Change of dead point produces two types of the same size, an opening type and a closing type. Setting it on the PP11-type pick and place unit makes grasping, lifting, rotating, chopping, and releasing possible with one electromagnetic valve.
2. Easy to use and reliable because of its simple mechanism and small size.
3. Arm working range is 30 mm up and down, and 90° and 180° left and right, respectively; speed is 60 mm/s up and down and 90°/s right and left.
4. Arm length is 200 mm maximum.

FREE TYPE: PICKING

PICK AND PLACE UNIT 2

Nakamura-kiki Engineering

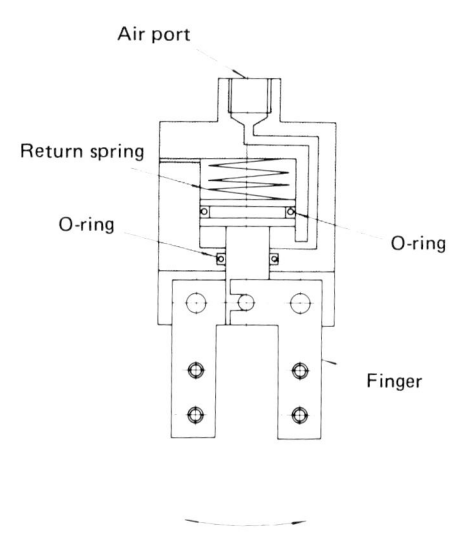

Working Speed			Working Range		
finger	wrist		finger	wrist	
open & shut	up & down	rotation	open & shut	up & down	rotation
0.3 s	Freely settable		19°	Freely designable	

Power Source	Sensors	Maximum Load	Size	Weight
Air pressure	None	1 kg	800 (L) × 120 (W) × 200 (H) mm	25 kg

NOTE

1. Two types of the same size corresponding to different air paths, an opening type and a closing type. Setting it on the PP01-type pick and place unit makes grasping, lifting, advancing, dropping, and releasing possible with one electromagnetic valve's on-off control.
2. Easy to use and reliable because of its simple mechanism and small size.
3. Arm working range is 30 mm up and down and 200 mm back and forth; speed is 60 mm/s up and down and 150 mm/s back and forth.

SLAVE HAND 1RL

Kayaba Industry Co., Ltd.

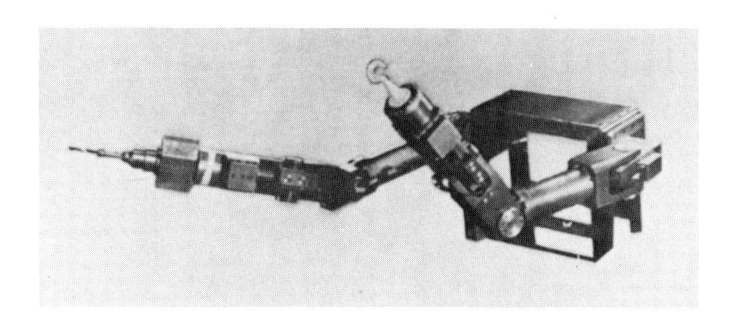

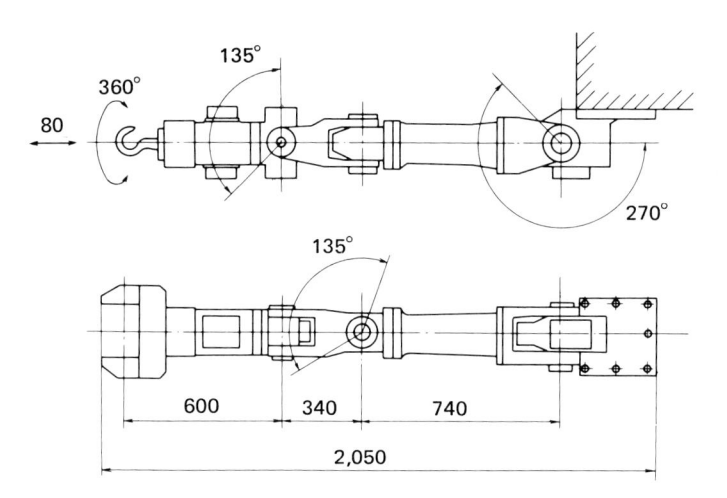

Working Speed				Working Range		
finger	wrist		finger	wrist	
open & shut	right & left	rotation	open & shut	right & left	rotation
5 mm/s at low	10 cm/s at low	7 rpm	100 mm	135°	360°
10 mm/s at high	15 cm/s at high	stretching			stretching
		30 mm/s			80 mm

Power Source	Sensors	Maximum Load	Size	Weight
Oil pressure	None	50 kg (for all positions), 1000 kg (vertically)	2050 mm	145 kg

NOTE

1. Can be worked simply at several hundred meters depth on the deep-sea bottom by a romote control from ship.
2. One-touch attachment of a special submarine tool to the hand tip.
3. Durable aluminum alloy used to lighten the submarine.
4. External picture of the hand is largely simplified and oil pipes and control valves are put inside the hand to enhance TV camera monitoring.
5. On-off control by joystick.
6. One-touch exchange of the tip tool.
7. Submarine operation possible.

FREE TYPE: PICKING

SLAVE HAND 2R

Kayaba Industry Co., Ltd.

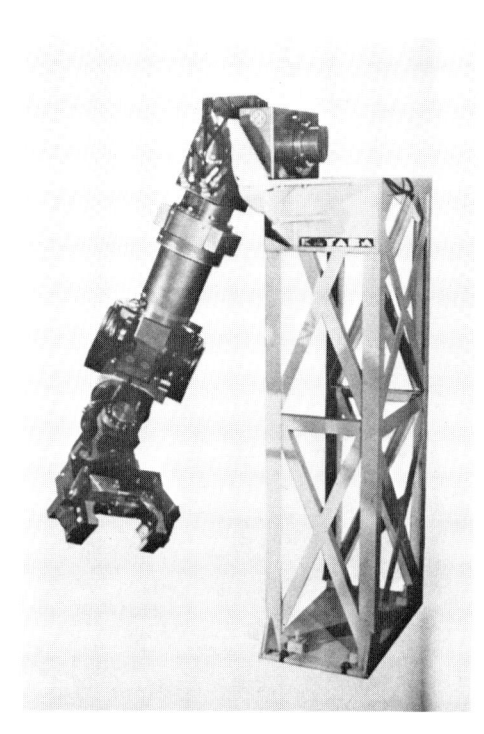

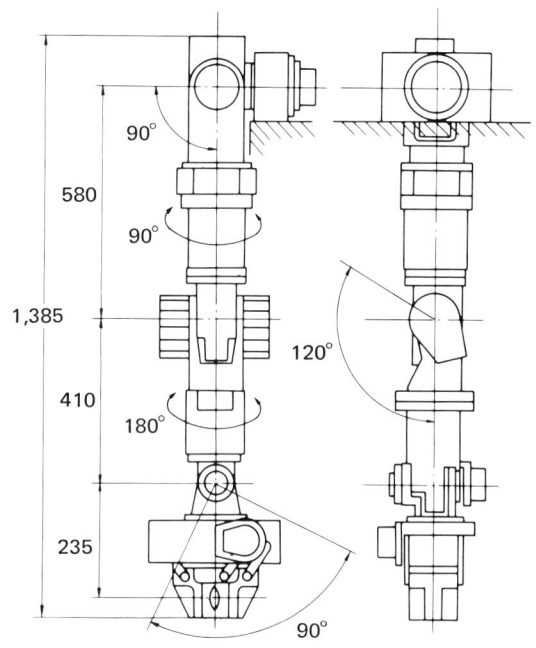

Working Speed				Working Range			
finger	wrist			finger	wrist	
open & shut	bending	rotation		open & shut	bending	rotation
120 mm/s	90°/s	180°/s		150 mm	90°	180°
Power Source	Sensors		Maximum Load	Size		Weight
Oil pressure	Force feedback		30 kg for all positions, 120 kg vertically	1385 mm		95 kg

NOTE

1. Each link of the hand has a force sensor, making complicated work possible at several hundred meters depth on the deep-sea bottom by remote control from ship.
2. Durable aluminum alloy used to lighten the submarine.
3. External picture of the hand is largely simplified to enhance TV camera monitoring.
4. Working range of the arm is 90° back-forth, right-left, and in rotation, and 120° for bend; speeds are 10°/s, 60°/s, and 20°/s, respectively.
5. Bilateral servo.
6. Master-slave type manipulator with oil pressure servo valves (submarine operation possible).

| HAND FOR LOADING AND UNLOADING | Kayaba Industry Co., Ltd. |

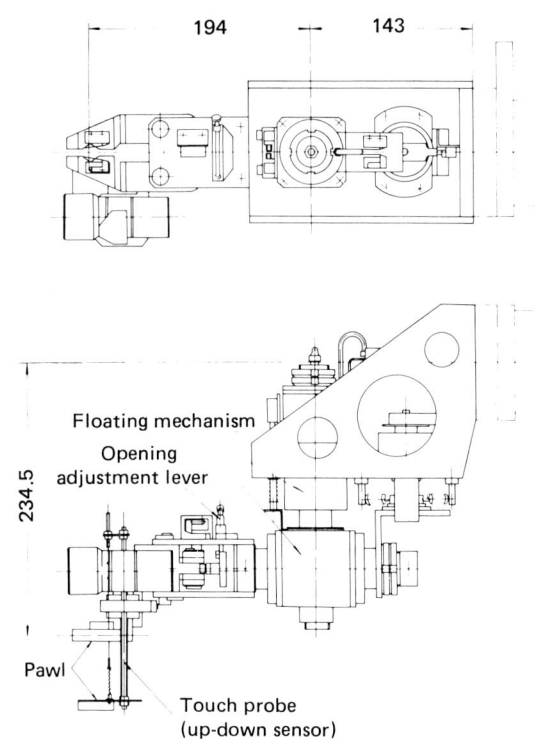

Working Speed				Working Range			
finger	wrist			finger	wrist		
open & shut	up & down		rotation	open & shut	up & down		rotation
Free	Freely designable			3-step opening	Freely designable		

Power Source	Sensors	Maximum Load	Size	Weight
Oil pressure	Touch probe	40 kg (max.)	337 × 234.5 mm	15 kg

NOTE

1. AC–DC servomotor control; value set by computer.
2. Combination of hand and touch probe (up-and-down axis sensor) to improve adaptability to various kinds of work and shorten the cycle time.
3. Floating mechanism (equipped with lock-unlock mechanism) for escape in grasping work.
4. Opening adjustment mechanism with a lever for grasping various kinds of work.
5. Maximum working range is 400 mm up-down, 800 mm back-forth, and 650 mm right-left; speeds are 500 mm/s, 700 mm/s, and 700 mm/s, respectively.

FREE TYPE: PICKING

| OIL PRESSURE HAND WITH SCREWING | Marol Company, Ltd. |

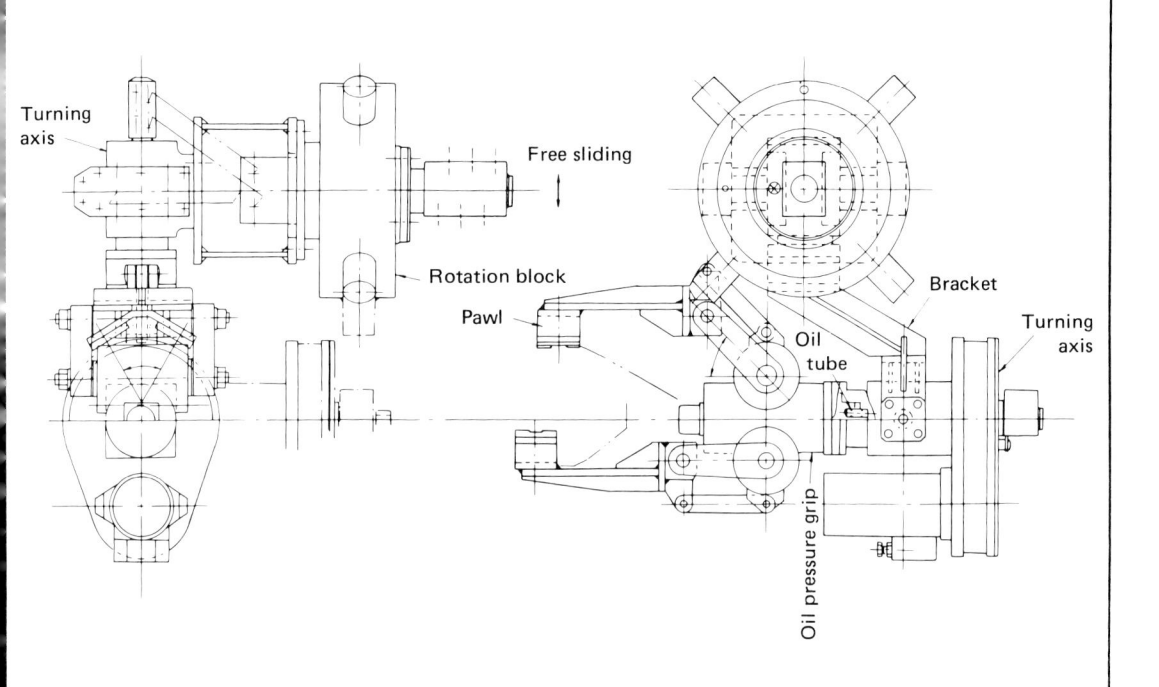

Labels on diagram: Turning axis, Free sliding, Rotation block, Pawl, Bracket, Oil tube, Turning axis, Oil pressure grip

Working Speed				Working Range			
finger		wrist			finger	wrist	
open & shut		up & down	rotation	open & shut	up & down	rotation
Freely adjustable		Freely adjustable		295 mm	20 mm	90°

Power Source	Sensors	Maximum Load	Size	Weight
Oil pressure	None	30 kg		185 kg

NOTE

1. Screwing into a nut with handling.
2. The work is a plug for seamless steel pipe manufacture.
3. Control: hand in shut, neutral, and open positions via an electromagnetic valve.
4. Grasping force: 43.7 kg.
5. Size range of work: 105 × 158 mm to 200.5 × 295 mm.

OIL PRESSURE HAND FOR CONICAL STEEL MATERIAL

Marol Company, Ltd.

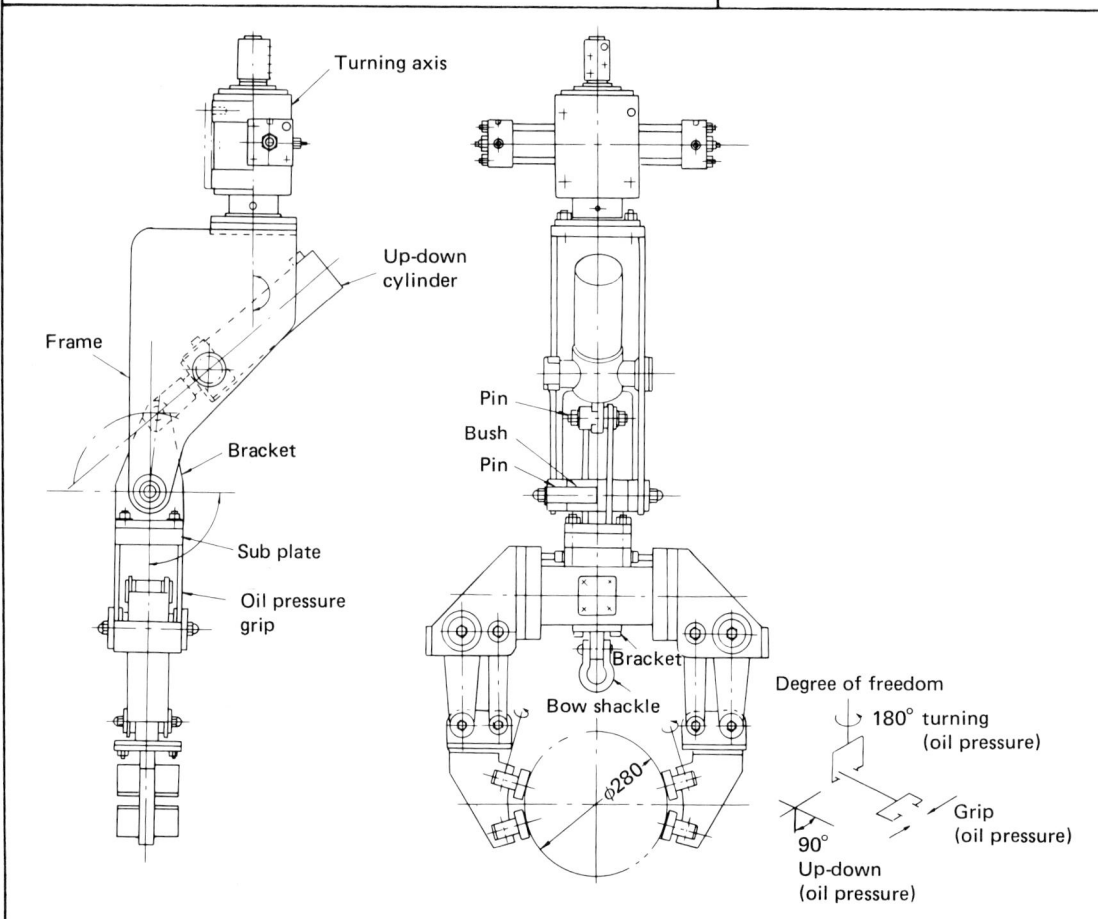

Working Speed				Working Range			
finger	wrist			finger	wrist	
open & shut	up & down	rotation		open & shut	up & down	rotation
Freely adjustable	5 s/90°	4 s/90°		170–300 mm	90°	180°

Power Source	Sensors	Maximum Load	Size	Weight
Oil pressure	None	350 kg		200 kg

NOTE

1. Used in handling conical steel material. The conical angle is about 25°, the size of the work is 190–280 mmφ or 318–450 mm.
2. Hand cylinder's push force is 3768 kg. Gripping force is 1150 kg.

FREE TYPE: PICKING

HAND FOR FUNNELS

Tokyo Keiki Co., Ltd.

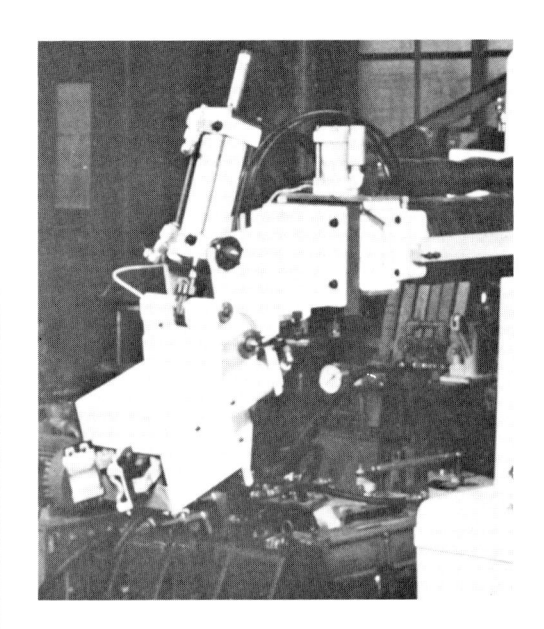

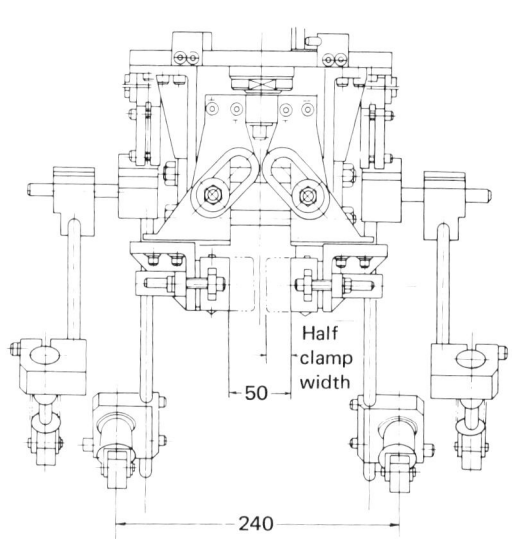

Half clamp width — 50
240

Working Speed			Working Range		
finger	wrist		finger	wrist	
open & shut	up & down	rotation	open & shut	up & down	rotation
0.5 s/40 mm	60°/s	90°/s	40 mm	Freely designable	

Power Source	Sensors	Maximum Load	Size	Weight
Oil pressure and air pressure	Touch sensor	10 kg		30 kg

NOTE

1. Two-position control by air pressure.

GRIP HAND FOR VERTICAL PICKING AND HORIZONTAL SETTING

Tokyo Keiki Co., Ltd.

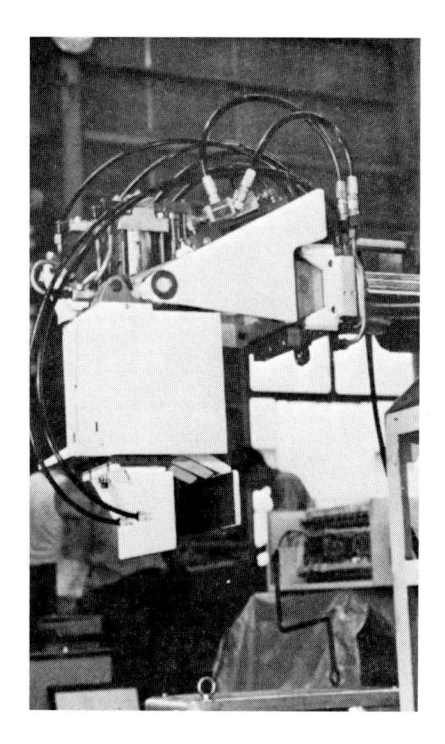

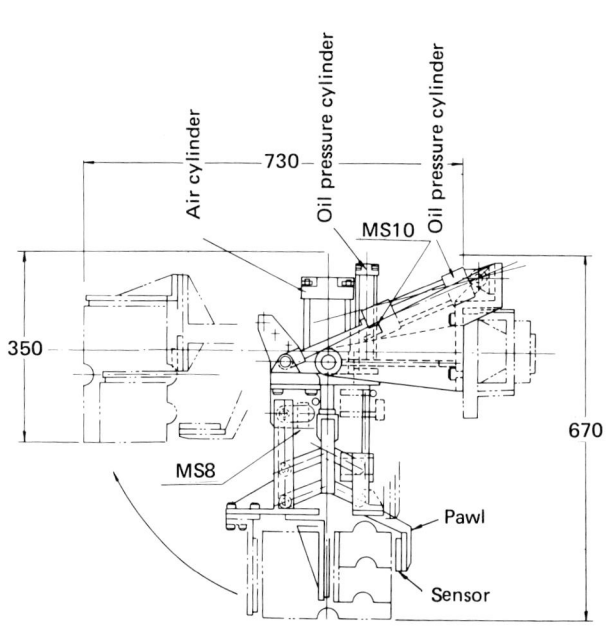

Working Speed			Working Range		
finger	wrist		finger	wrist	
open & shut	up & down	rotation	open & shut	up & down	rotation
0.5 s/40 mm	60°/s		40 mm	90°	

Power Source	Sensors	Maximum Load	Size	Weight
Oil pressure and air pressure	Touch sensor	30 kg	670 × 350 mm	35 kg

NOTE

1. Three methods of control used together: electricity-oil pressure two-position control, air pressure two-position control, and vacuum absorption two-position control.

FREE TYPE: PICKING

MAY HAND MIII

Yamatake Honeywell Co., Ltd

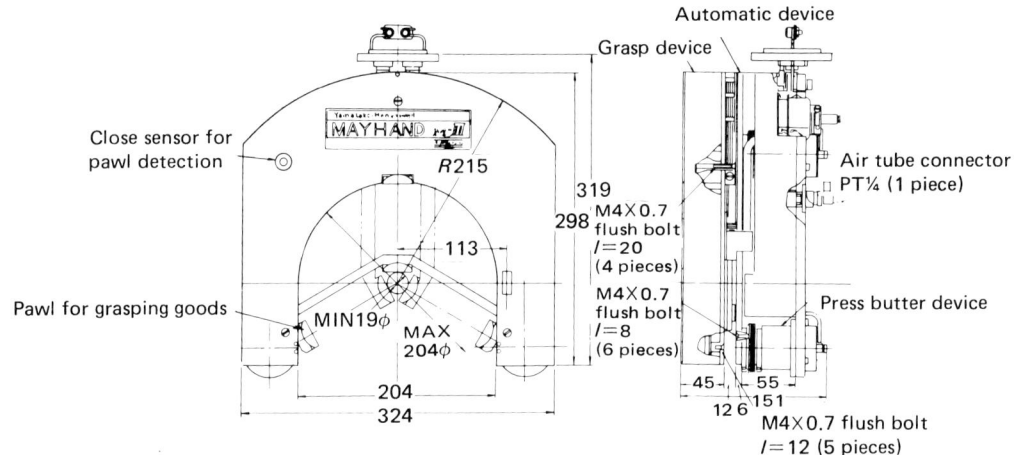

Working Speed			Working Range		
finger	wrist		finger	wrist	
open & shut	up & down	rotation	open & shut	up & down	rotation
40 mm/s	90°/s		20-200 mmφ	160°	

Power Source	Sensors	Maximum Load	Size	Weight
Air pressure and electricity	Grasping force sensor (from the torque of a reduction gear)	10 kg	324 (W) × 298 (H) × 151 (T) mm	12 kg

NOTE

1. Makes accurate holding possible for highly precise mechanical processing; that is, the eccentric angle can be canceled and the material grasped and centered on the chuck face.
2. Enables 0.01-mm mechanical processing.
3. Work center's offset is small because of synchronization of three pawls.
4. Range of work is from φ20 to φ204 and wide; internal grasping is also possible.
5. The pawl position can be detected because it rotates around the work as a center.
6. Control: position control (on–off servo) and force control (on–off).

CHUO HAND X2	Chuo University

Working Speed		Working Range	
finger	wrist	finger	wrist
open & shut	up & down rotation	open & shut	up & down rotation
	Freely settable	50 mm	280° 360°

Power Source	Sensors	Maximum Load	Size	Weight
Electricity (AC–DC servomotors)	Strain gauges, on–off switches		Arm total length 935 mm	

NOTE
1. Designed for study of control and robot language of multijoint manipulator with vision; controlled by a PDP 8/e minicomputer (16 Kword).
2. Control by AC–DC servomotors.
3. Working range of arm is 110° up-down, 300° for internal or external rotation, and 180° right-left.

FREE TYPE: SNAP   77

| U.I. HAND | University of Tokyo |

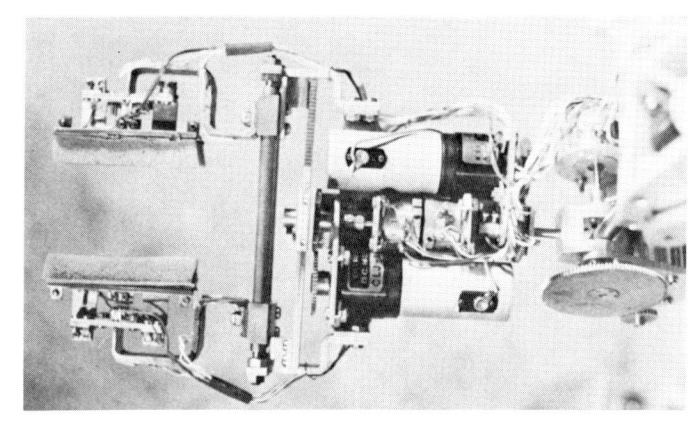

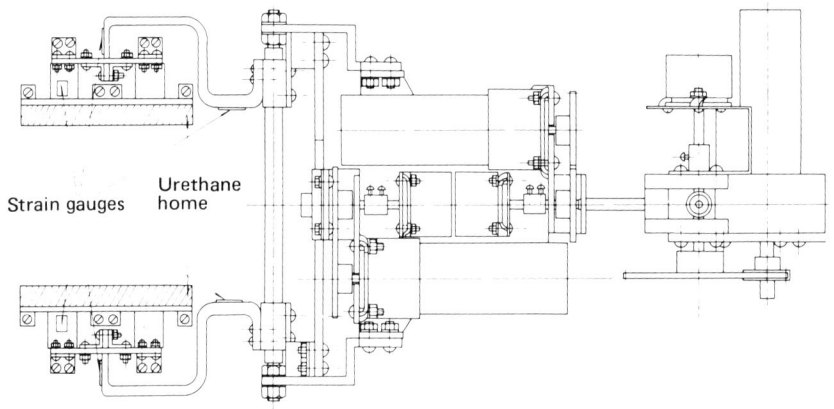

Strain gauges — Urethane home

Working Speed			Working Range		
finger	wrist		finger	wrist	
open & shut	right & left, up & down	rotation	open & shut	right & left, up & down	rotation
About 10 mm/s	10°/s for each		0-85 mm	$-150°-(+150)°$ / $-60°-(+30)°$	$-150°-(+150)°$

Power Source	Sensors	Maximum Load	Size	Weight
Electricity (motors)	Strain gauges	500 g	234 × 132 mm (hand) 530 × 1100 × 720 mm (body)	2 kg (hand) 32 kg (body)

| NOTE |

1. Tip position and attitude degrees of freedom are mechanically separated; the transformation from joint angles for the tip position and attitude to the reverse is simplified.
2. Torque induced by its own weight is eliminated by two balance weights; the servo has no load in any case whenever it has nothing.
3. Force sensor installed at the tip picks up the external force of the tip and grasping force, and force differences of no more than 10 g can be detected.

TDL DUAL HAND

Tokyo Denki University

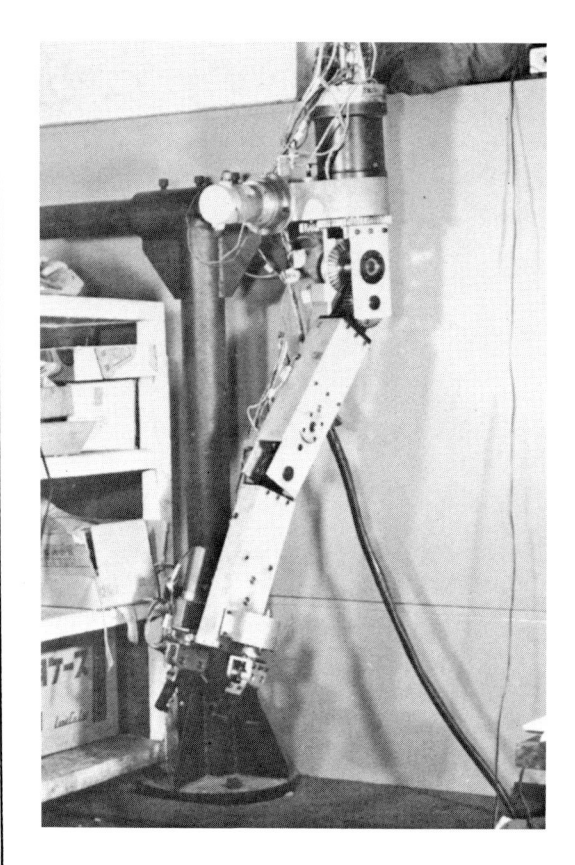

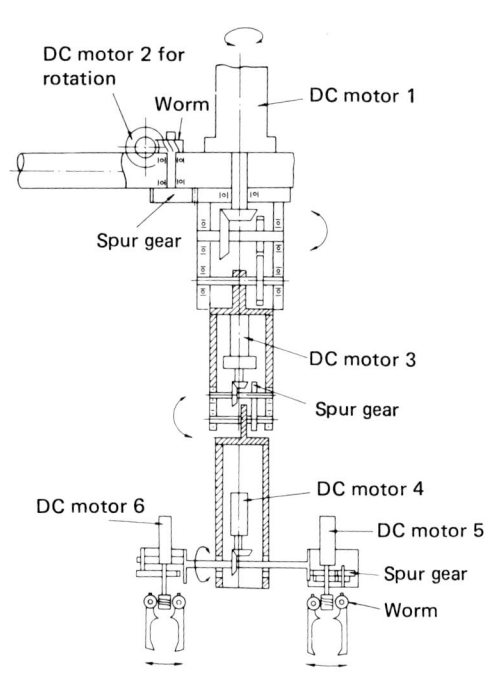

Working Speed			Working Range		
finger	wrist		finger	wrist	
open & shut	up & down	rotation	open & shut	up & down	rotation
0.3 s/90°	0.6 s/90°		Free	Freely settable	

Power Source	Sensors	Maximum Load	Size	Weight
Electricity (DC motors)	None	1 kg	1.2 m	About 40 kg

NOTE

1. Point-to-point control method.
2. Encoders used for motors 1 and 2, potentiometers for motors 3–6.
3. Pinboard order method.
4. IC 200-W amplifier.
5. Working speed of the arm is 90°/2 s up-down, 90°/1.5 s back-forth, and 90°/2.5 s right-left.

FREE TYPE: SNAP

HIGH-QUALITY MANIPULATOR

Mechanical Engineering Laboratory

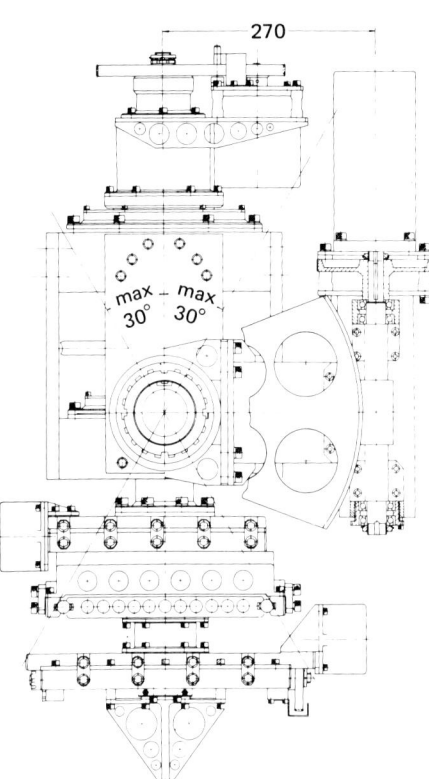

Working Speed		Working Range		
finger	wrist || finger	wrist	
open & shut	up & down rotation || open & shut	up & down rotation	
Freely settable	Settable manually or by computer || 100 mm	10 mm	10 mm, 360°
Power Source	Sensors || Maximum Load	Size	Weight
Electricity (pulse motors)	Vision, fiberscope, and visicon || 2 kg	Total length 1100 mm	250 kg

NOTE

1. Controlled by computer.
2. Order signal is pulse, and tip displacement angle compensation per one pulse for each degree.
3. Working range of the arm is 100 mm up-down, 60° for rotation, and 60° for twist.

ETL MANIPULATOR

Electrotechnical Laboratory

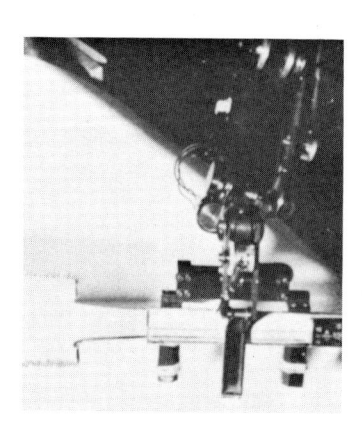

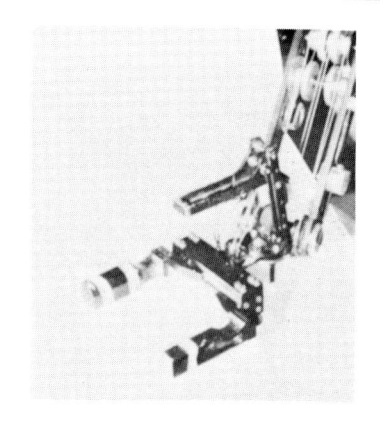

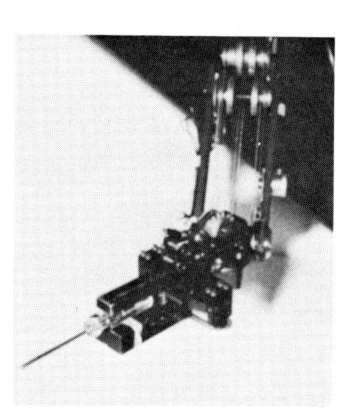

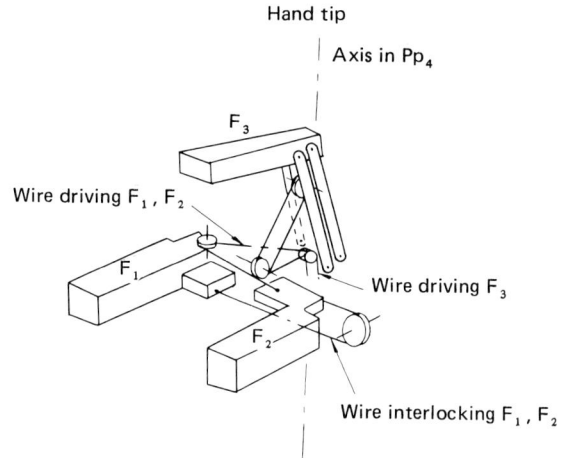

Hand tip
Axis in Pp_4
F_3
Wire driving F_1, F_2
F_1
F_2
Wire driving F_3
Wire interlocking F_1, F_2

Holding a driver

Working Speed			Working Range		
finger	wrist		finger	wrist	
open & shut	up & down	rotation	open & shut	up & down	rotation
	3 rad/s	3 rad/s	0–100 mm for parallel two fingers, 80 mm for the third one	±90°	±90°

Power Source	Sensors	Maximum Load	Size	Weight
Electricity	None	1 kg	9° × 45 cm (arm), 8° × 45 cm (forearm), 170 mm (L) × 140 mm (W) (hand)	

NOTE

1. Software servo, and open-loop control of force, making reduction ratio of actuators low and retaining reversibility. The arm has 7 + 2 degrees of freedom and control with redundancy is possible.
2. Each joint is driven by adjusting the tension of two wires opposed to each other.
3. Magneto-fine powder clutch, such that the converted torque is proportional to the applied current, is introduced as an actuator as well as the usual ones.
4. Direct computer software servo method. Angle (potentiometer) and angular velocity (tachometer) are input into the computer through an analog-to-digital converter and the output of torque controls the magneto-fine clutch.

FREE TYPE: SNAP

EXTERNAL GRASPING-TYPE HAND WITH FLOATING SWING LEVERS

Yasukawa Electric Mfg. Co., Ltd.

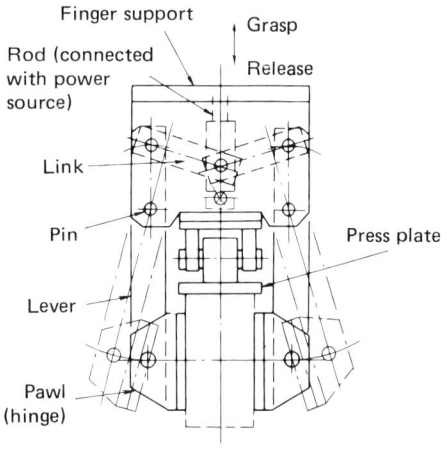

Finger support — Grasp
Rod (connected with power source) — Release
Link
Pin
Press plate
Lever
Pawl (hinge)

Working Speed				Working Range			
finger			wrist		finger		wrist
open & shut			up & down	rotation	open & shut		up & down	rotation
1 s [STD timer 1.5 s]			Freely settable		Free		Freely designable	

Power Source	Sensors	Maximum Load	Size	Weight
Electricity (motor)	None	About 40 kg	About 300 × 150 mm (work)	About 36 kg

NOTE

1. If the work width has dispersion, the pawl floats to face the work.
2. Descent of the loader is stopped by a limit switch which recognizes the upward slide of the pawl part, the work-holding plate hitting the work.
3. On-off control by a timer and a limit switch.

	Yasukawa Electric Mfg. Co., Ltd.
EXTERNAL GRASPING AND HOLDING HAND	

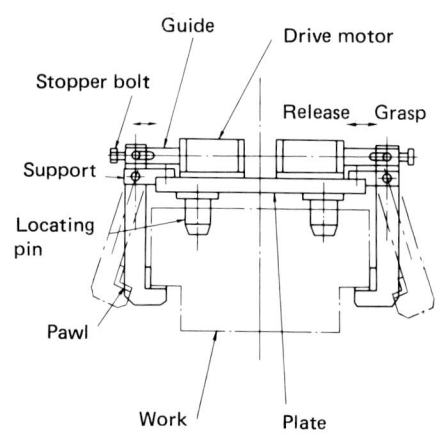

Diagram labels: Guide, Drive motor, Stopper bolt, Release, Grasp, Support, Locating pin, Pawl, Work, Plate

Working Speed			Working Range	
finger	wrist		finger	wrist
open & shut	up & down	rotation	open & shut	up & down	rotation
1 s [STD timer 1.5 s]	Freely settable		Free	Freely designable

Power Source	Sensors	Maximum Load	Size	Weight
Electricity (motors)	None	About 140 kg	About 400 × 250 mm (work)	About 28 kg

NOTE

1. Two power sources are used. Each pawl is opened independently because of the width and weight of the work and the accuracy of the work position. The position pins inserted into the basic holes prevent error.
2. On–off control by a timer and a limit switch.

FREE TYPE: SNAP

EXTERNAL GRASPING AND HOLDING HAND	Yasukawa Electric Mfg. Co., Ltd.

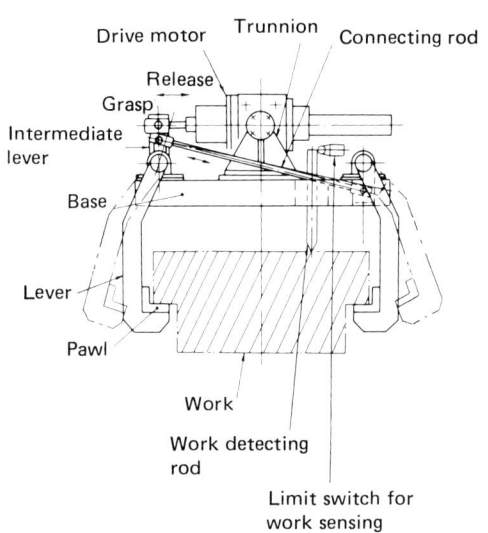

Diagram labels: Drive motor, Trunnion, Connecting rod, Release, Grasp, Intermediate lever, Base, Lever, Pawl, Work, Work detecting rod, Limit switch for work sensing

Working Speed				Working Range		
finger	wrist		finger	wrist	
open & shut	up & down	rotation	open & shut	up & down	rotation
1 s	Freely settable		Free	Freely designable	

Power Source	Sensors	Maximum Load	Size	Weight
Electricity (motors)	None	About 400 kg	About 1000 × 1000 mm	About 150 kg

NOTE
1. The pawls must be wide because of the width and weight of the work.
2. Four pawls are controlled through links and rods by one power source.
3. On-off control by a limit switch.

MEL ARM	Mechanical Engineering Laboratory

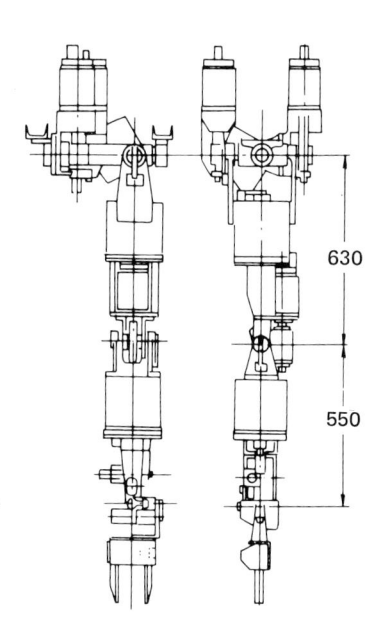

Working Speed		Working Range		
finger	wrist		finger	wrist	
open & shut	up & down	rotation	open & shut	up & down	rotation
	7.5°/s	31°/s	0–650 mm	+14°–(−45)° +90°–(−45)°	+150°–(−150)°

Power Source	Sensors	Maximum Load	Size	Weight
Electricity, oil pressure	Force sense, touch sense	10 kg	300 mm in length	100 kg (each arm), 3 kg (hand)

NOTE
1. Human arm-type joint structure.
2. Cooperative control of two arms.
3. Manipulation by a master manipulator.
4. Manipulation by each joystick of the wrist and the hand.
5. Minicomputer control.
6. Arm and forearm weigh 81 and 18 kg, respectively. Shoulder has 5 degrees of freedom, that is, back-forth, right-left, rotation, and bending the elbow; wrist has 3 degrees of freedom, rotation, up-down, and swinging.

FREE TYPE: SNAP

ARTIFICIAL HAND AND ARM CONTROLLED BY ELECTRICITY AND OIL PRESSURE

Kyushu University

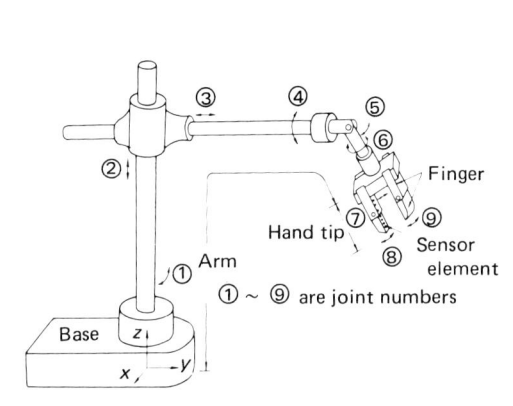

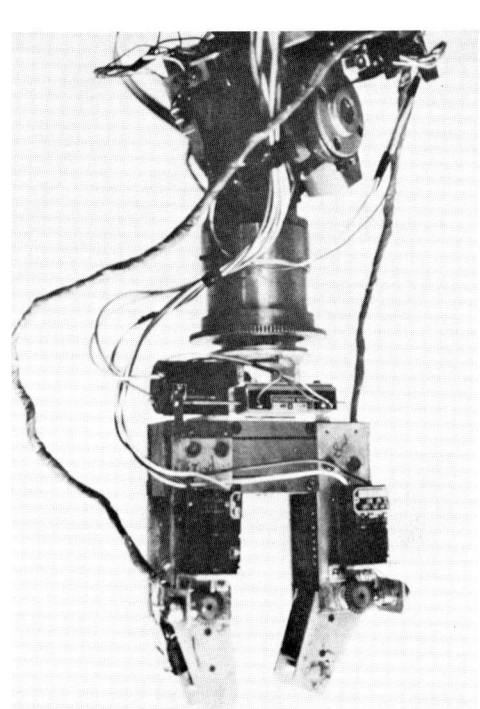

Working Speed				Working Range			
finger	wrist			finger	wrist	
open & shut	up & down	rotation		open & shut	up & down	rotation
30 mm/s (max.)	1/2 rad/s (max.)	1/2 rad/s (max.)		80 mm	−1/2−1/6 rad	±1/2 rad

Power Source	Sensors	Maximum Load	Size	Weight
Electricity, oil pressure	Pressure sensors, strain gauges	1 kg	200 (L) × 700 (W) × 700 (H) mm	400 kg (except oil pressure source)

NOTE

1. Our laboratory has made the hand tip; the hand base and arm are UNIMAN, an industrial robot made by Fujikoshi.
2. Artificial hand and arm are controlled by electricity and oil pressure; the hand tip is controlled by electricity.
3. If there is an obstacle in the working space, the arm can avoid the obstacle and move itself to an object.
4. Pressure sensors at the finger make it possible to recognize the shape, size, and material of the grasped object.

MULTIJOINTED ROBOT HAND (TYPE KAR-2)

Kayaba Industry Co., Ltd.

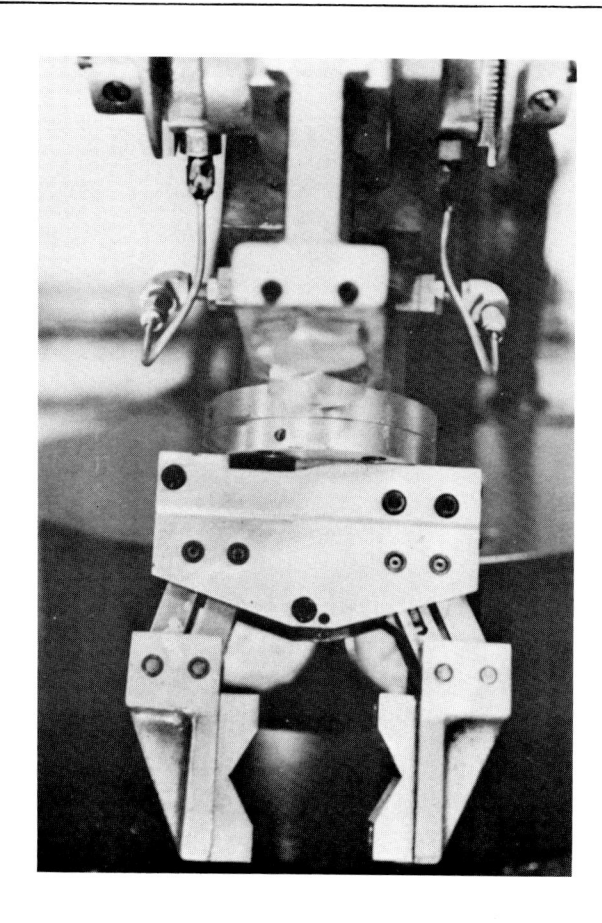

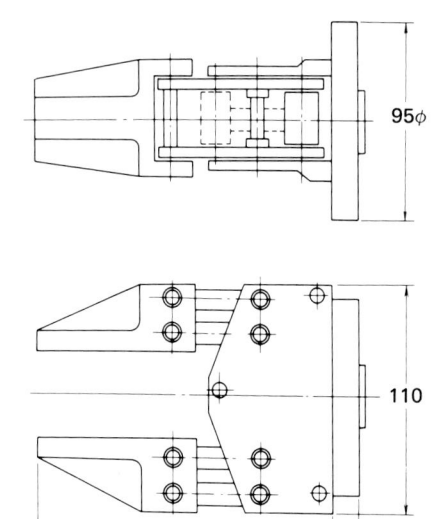

	Working Speed			Working Range	
finger	wrist		finger	wrist	
open & shut	up & down	rotation	open & shut	up & down	rotation
Freely settable	60°/s	60°/s	Freely designable	180°	180°

Power Source	Sensors	Maximum Load	Size	Weight
Oil pressure (140 kg/cm²), electricity (AC 200 V, 3.7 kW)	None	12 kg (max.)	1000 (L) × 600 (W) × 1575 (H) mm	400 kg

NOTE

1. Working range of the arm is 125° up-down, 240° right-left, and 150° back-forth; speed is 30°/s up-down, 90°/s right-left, and 45°/s back-forth.
   - Operation: point-to-point control
   - Position detection: encorder
   - Sequence memory: magnetic drum
   - Memory size: magnetic drum
   - Position memory: 256 in standard
   - Position setting: electricity-oil pressure servo
   - External synchronizing signal: 4 receiving × 4 transmitting
   - Position setting accuracy: ±1 mm

FREE TYPE: SNAP

HAND FOR MARINE WORK

Marine Science & Technology Center

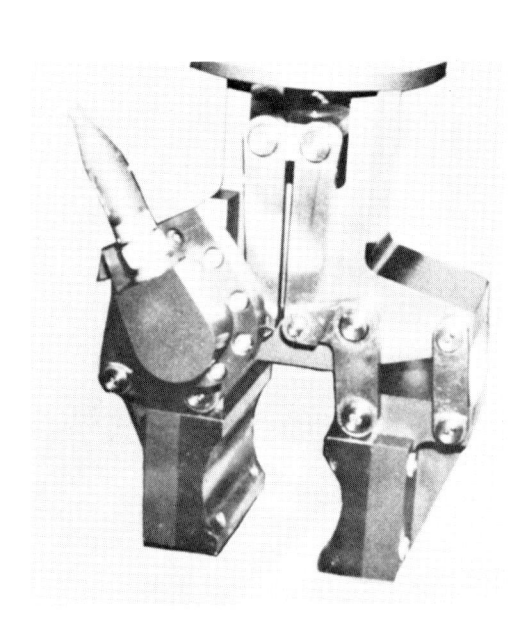

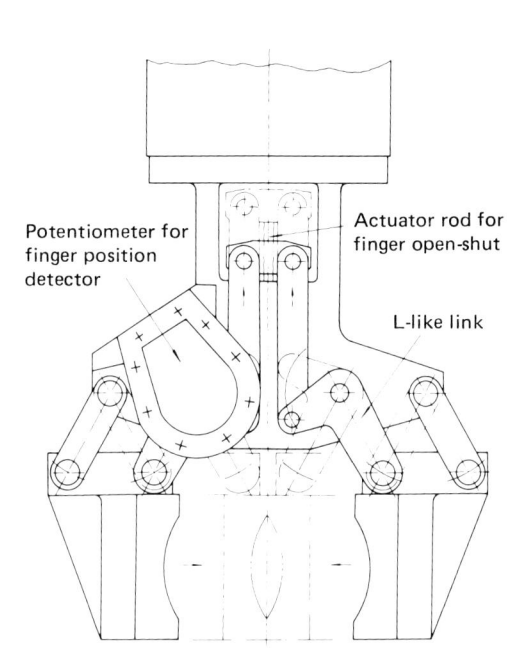

Labels on diagram:
- Potentiometer for finger position detector
- Actuator rod for finger open-shut
- L-like link

Working Speed				Working Range		
finger	wrist		finger	wrist	
open & shut	bend	rotation	open & shut	up & down	rotation
100 mm/2 s	90°/s (right hand)	180°/2 s			

Power Source	Sensors	Maximum Load	Size	Weight
Oil pressure (200 kg/cm² max.)	Force sensor	30 kgf (right hand), 50 kgf (left hand)	1385 mm total right hand, 1430 mm total left hand	

NOTE

1. Uses master-slave method developed for marine work.
2. Operator can detect constraining force acting on the manipulator by using bilateral servo method.
3. Electricity-oil pressure servomechanism.
4. Hardness of the grasped body can be determined. Maximum sensitivity is about 100 kgf.
5. Working speeds are 120°/3 s (right arm) and 120°/4 s (left arm) for elbow bending, 90°/3 s (right arm) and 90°/4 s (left arm) for back-forth or right-left swing at the shoulder.

MEL HAND

Mechanical Engineering Laboratory

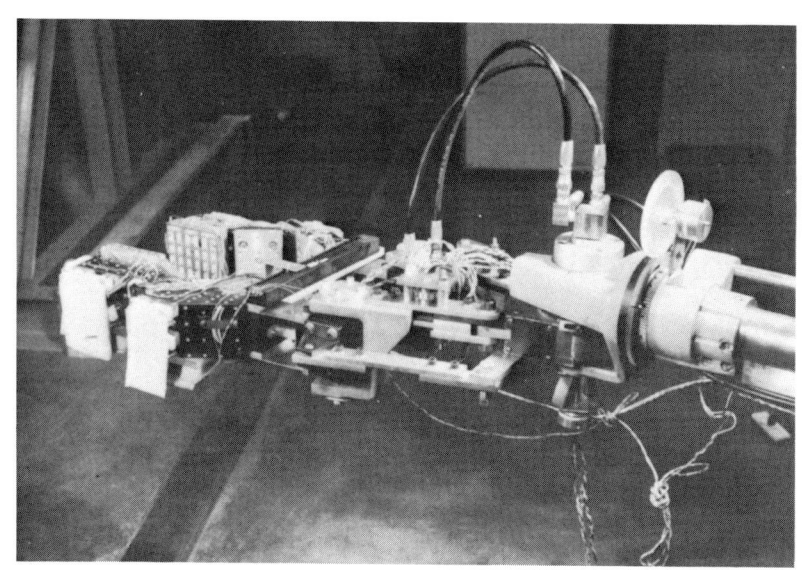

Labels on drawing: Slide ball bearing; Coil spring; Linear displacement type potentiometer; Coil spring

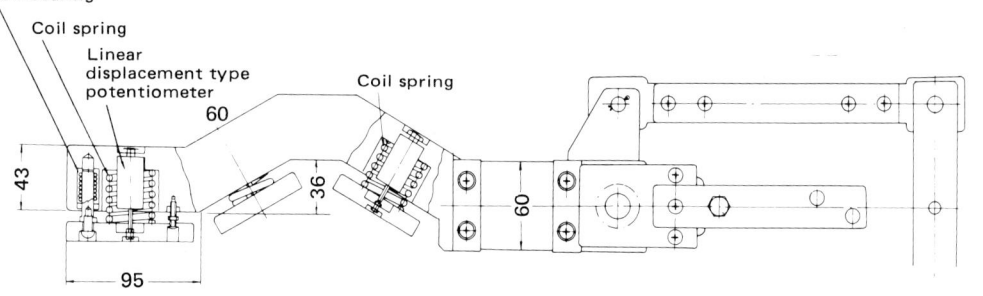

Dimensions: 43, 60, 36, 60, 95

Working Speed				Working Range		
finger	wrist		finger	wrist	
open & shut	up & down	rotation	open & shut	up & down	rotation
Two-position settable			130 mm	0–180°	0–180°

Power Source	Sensors	Maximum Load	Size	Weight
Oil pressure	Touch sensor: on–off switches; force sensor: springs and potentiometers	15 kg	400 mm	15 kg

NOTE

1. Controlled by computer.
2. Three degrees of freedom of the arm have a rectilinear compensation circuit as a hand.
3. Weight is detected by both a leaf spring and a differential transformer.
4. Working range of the arm is 760 mm up-down and back-forth, 0–240° right-left.

FREE TYPE: SNAP

OIL PRESSURE GRIP | Marol Company, Ltd.

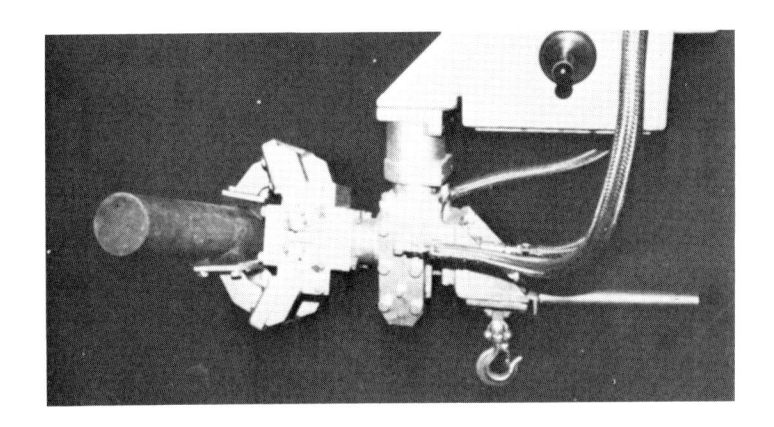

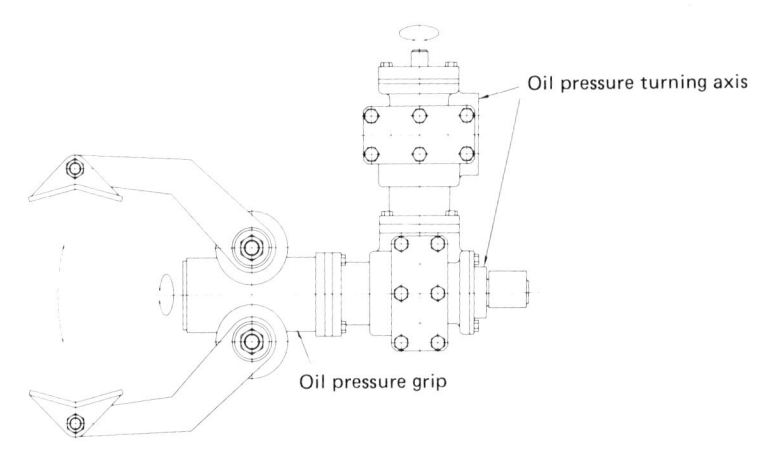

Oil pressure turning axis

Oil pressure grip

Working Speed			Working Range		
finger	wrist		finger	wrist	
open & shut	up & down	rotation	open & shut	up & down	rotation
Freely adjustable	Freely adjustable		Designable to order	Dependent on arm order	±90°

Power Source	Sensors	Maximum Load	Size	Weight
Electricity, oil pressure	None	100–500 kg		70–200 kg except for fingers

NOTE

1. Electromagnetic valve switching method.
2. Flow rate adjusted in accordance with the work by a variable throttle valve to limit the maximum working speed.

OIL PRESSURE GRIP	Marol Company, Ltd.

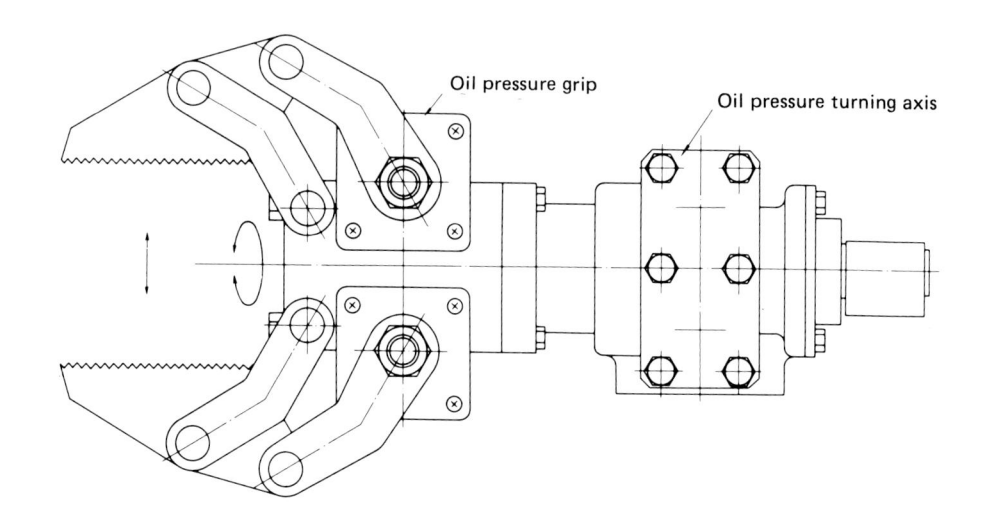

Oil pressure grip

Oil pressure turning axis

Working Speed		Working Range		
finger	wrist		finger	wrist	
open & shut	up & down	rotation	open & shut	up & down	rotation
Freely adjustable	Freely adjustable		Designed to order	Dependent on arm order	±90°

Power Source	Sensors	Maximum Load	Size	Weight
Electricity, oil pressure	None	100–600 kg		50-100 kg except for fingers

NOTE

1. Electromagnetic valve switching method.
2. Flow rate adjusted in accordance with the work by a variable throttle valve to limit the maximum working speed.

FREE TYPE: SNAP

OIL PRESSURE HAND FOR COPPER WIRE BOBBINS

Marol Company, Ltd.

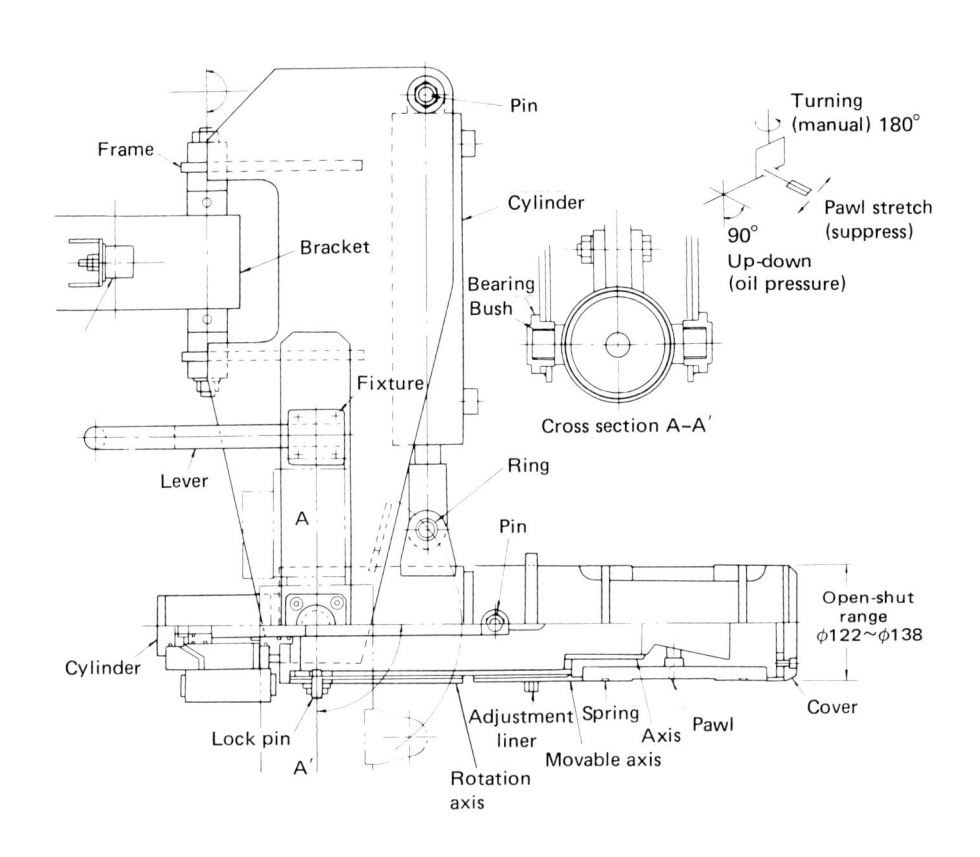

Working Speed			Working Range		
finger	wrist		finger	wrist	
open & shut	up & down	rotation	open & shut	up & down	rotation
Freely adjustable	Freely adjustable		122-138 mm	90°	180°

Power Source	Sensors	Maximum Load	Size	Weight
Oil pressure	None	150 kg		100 kg

NOTE

1. Made to handle copper wire bobbins.
2. 320 mm external diameter, 360 mm length, 135 mm chuck hole diameter.
3. Maximum chucking load 150 kg.

OIL PRESSURE HAND FOR WHEEL AXLES

Marol Company, Ltd.

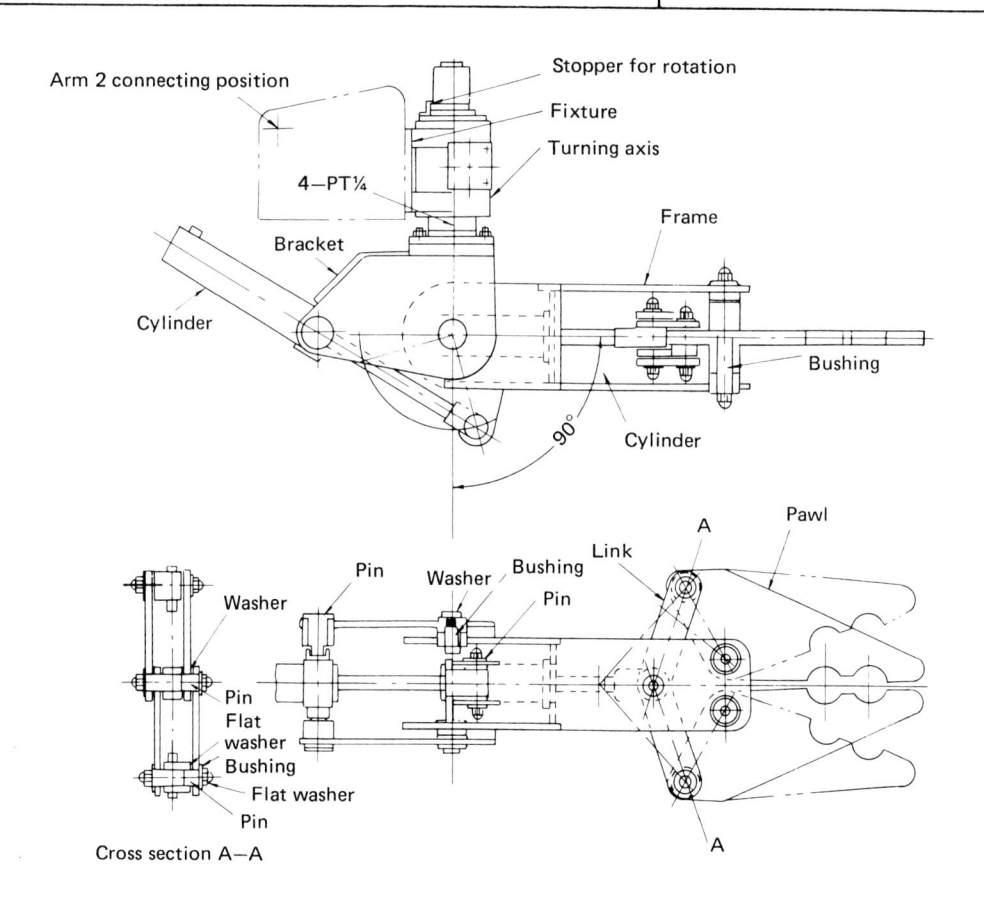

Working Speed			Working Range		
finger	wrist		finger	wrist	
open & shut	up & down	rotation	open & shut	up & down	rotation
Free	5 s/90°	4 s/90°	120 mm (max.)	90°	180°

Power Source	Sensors	Maximum Load	Size	Weight
Oil pressure	None	150 kg		160 kg

NOTE

1. Made to handle objects such as crankshafts, wheel axles, etc.
2. 760–1280 mm or 1410–1900 mm work length.
3. 70-kg m revolving torque.

FREE TYPE: SNAP

CLAMP HAND

Marol Company, Ltd.

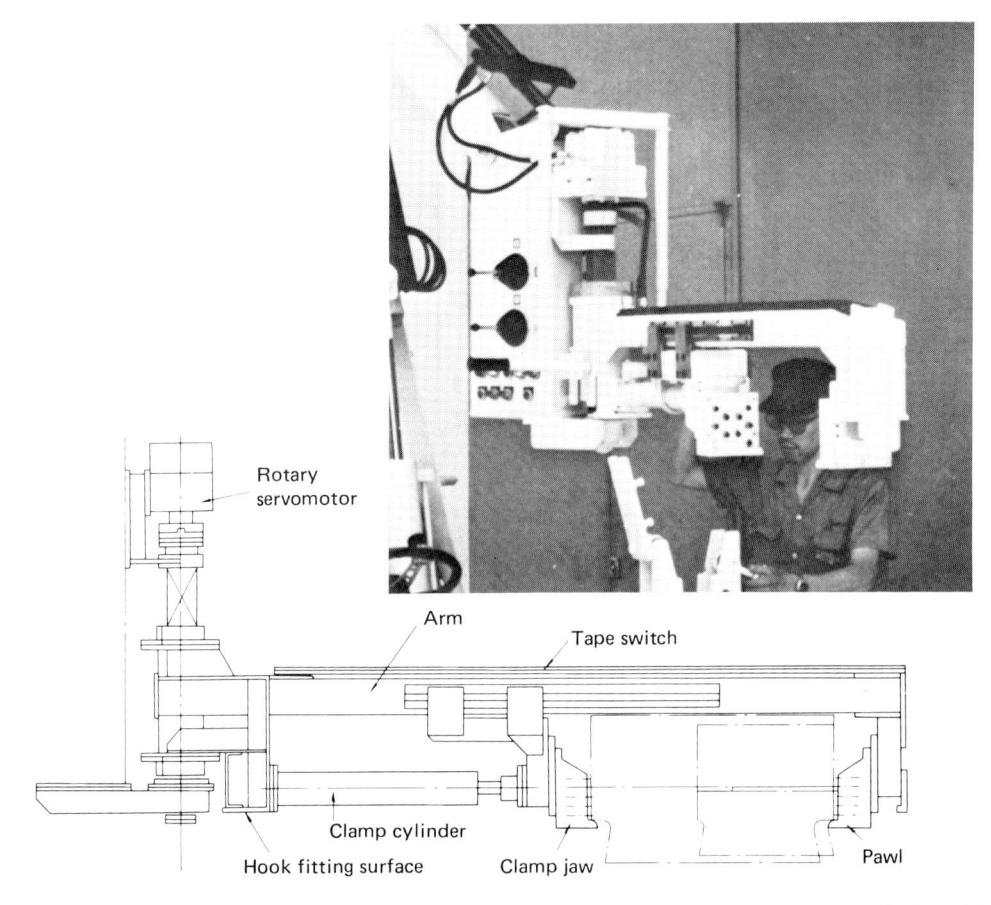

Rotary servomotor
Arm
Tape switch
Clamp cylinder
Hook fitting surface
Clamp jaw
Pawl

Working Speed			Working Range		
finger	wrist		finger	wrist	
open & shut	up & down	rotation	open & shut	up & down	rotation
Free	Freely settable		540 mm		±45°

Power Source	Sensors	Maximum Load	Size	Weight
Oil pressure	A tape switch	325 kg	600 (L) × 320 (H) × 150 (W) mm	180 kg

NOTE

1. Hand for press metal mold.
2. Driven by a rotary servomotor.
3. Has ten jaw pawls that move coherently. All pawls contact work simultaneously.
4. When the tape switches on, the arm departs automatically from the press machine and returns after a time set by a timer.

HAND FOR COLUMNS

Daikin Kogyo Co., Ltd.

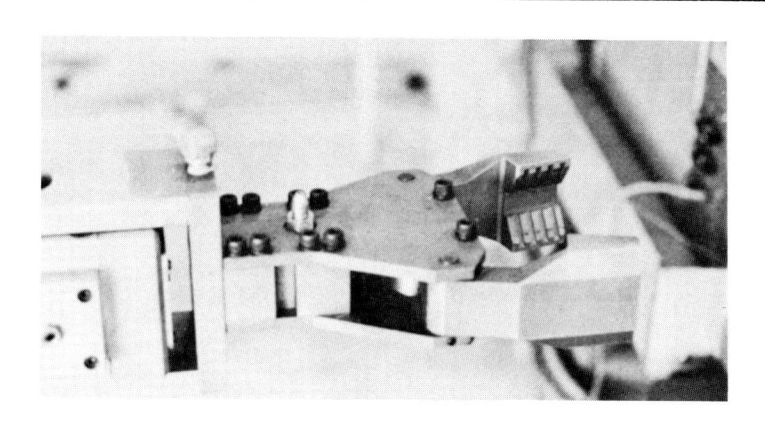

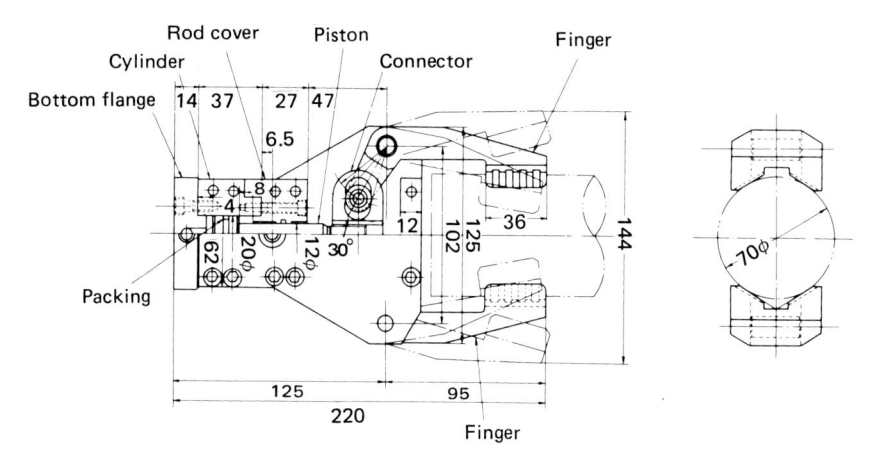

Labels on diagram: Rod cover, Piston, Finger, Cylinder, Connector, Bottom flange, Packing, Finger

Dimensions: 14, 37, 27, 47, 6.5, 4, 8, 62, 20φ, 12φ, 30°, 12, 36, 125, 102, 144, 125, 95, 220, 70φ

	Working Speed			Working Range	
finger	wrist		finger	wrist	
open & shut	swing	rotation	open & shut	up & down	rotation
0.4 s	120°/s	120°/s	125 mm	130°	180°

Power Source	Sensors	Maximum Load	Size	Weight
Oil pressure (70 kg/cm²)	None	10 kg (total robot weight)	220 × 144 mm	4.6 kg

NOTE

1. Has no connecting links and is simple, although equipped with a force amplification mechanism.
2. Bearing lubricating surfaces of the connecting parts are quenched.
3. Working range of the arm is 500 mm up-down, 300 mm back-forth, and 230° right-left; working speeds are 500 mm/s, 700 mm/s, and 110°/s, respectively.
4. Memory recovery type servo control.
5. On-off control in finger.
6. 22 kg grasping force.

FREE TYPE: SNAP

HAND FOR WEIGHT LIFTING INVERSION

Daikin Kogyo Co., Ltd.

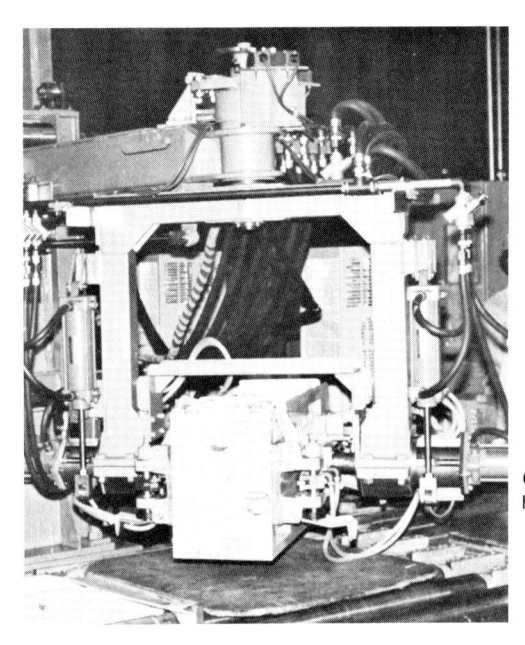

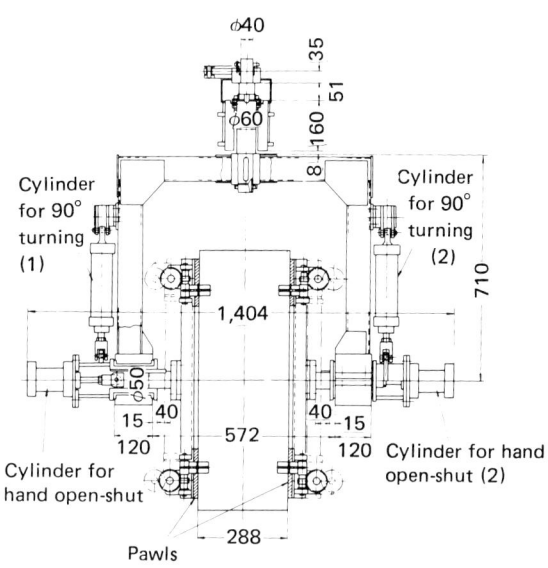

Working Speed		Working Range	
finger	wrist	finger	wrist
open & shut	up & down rotation	open & shut	up & down rotation
400 mm/0.5 s	Freely settable	80 mm	Freely designable

Power Source	Sensors	Maximum Load	Size	Weight
Oil pressure	None	50 kg	1404 × 700 mm	130 kg

NOTE

1. The large chucking stroke of the hand makes good control possible regardless of inexact work location.
2. Equipped with a 90° turning mechanism and a 45° rotating mechanism.
3. Oil pressure circuit for hand-opening cylinder has a pressure adjusting valve to prevent deformation or damage to the casing.

OIL PRESSURE HAND

Okamura Mfg. Co., Ltd.

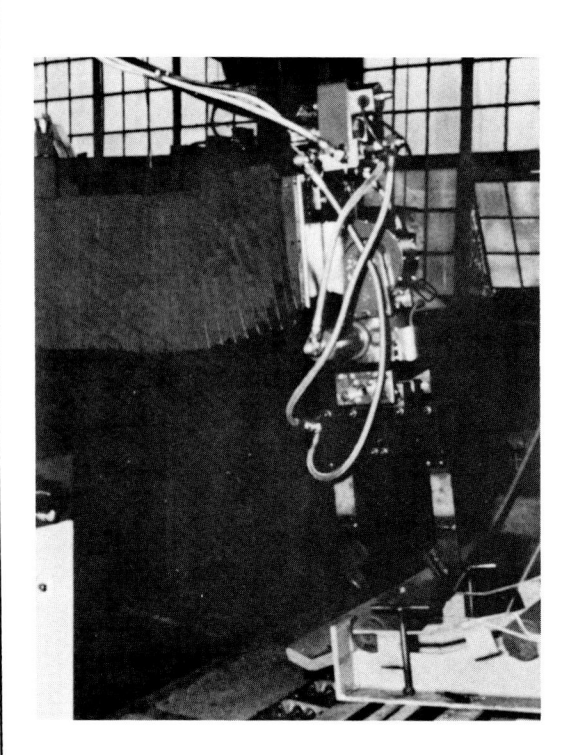

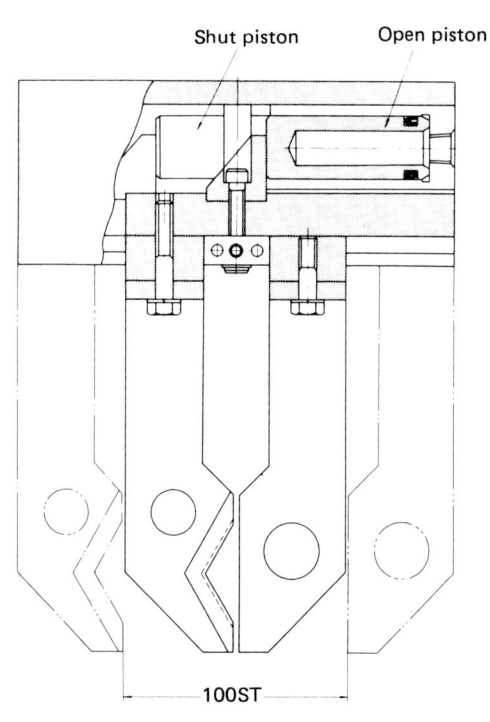

Shut piston Open piston

100ST

Working Speed			Working Range		
finger	wrist		finger	wrist	
open & shut	up & down	rotation	open & shut	up & down	rotation
0.3 s	Freely settable		100 mm (max.)	Freely designable	
Power Source	**Sensors**	**Maximum Load**	**Size**		**Weight**
Oil pressure	None	20 kg (max.)	100 × 200 × 80 mm (except for pawls)		

NOTE

1. Total external size is small for the open-and-shut length of 100 mm.
2. Clamp uncertainty is eliminated by three-pawl support.
3. In clamping objects such as rods, the center can be held constant even if the external diameter changes.

FREE TYPE: SNAP   97

STANDARD HAND FOR KAWASAKI UNIMATES | Kawasaki Heavy Industries, Ltd.

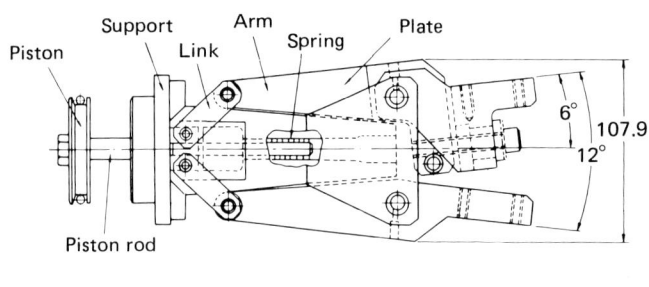

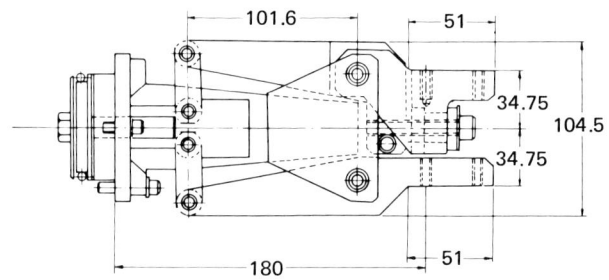

Working Speed				Working Range			
finger	wrist			finger	wrist		
open & shut	up & down		rotation	open & shut	up & down		rotation
Free	Freely settable			12°	Freely designable		

Power Source	Sensors	Maximum Load	Size	Weight
Air pressure (5 kg/cm² G)	None	About 30 kg	180 × 104 mm	3 kg

NOTE

1. Made as a standard hand for Kawasaki Unimates; the finger is exchangeable for different purposes in general handling.
2. Standard air cylinders and air circuit for robots.
3. Work clamp by toggle lock mechanism is reliable.
4. On-off control by air pressure.

PARALLEL MOVABLE FINGER — Seiko-Seiki Co., Ltd.

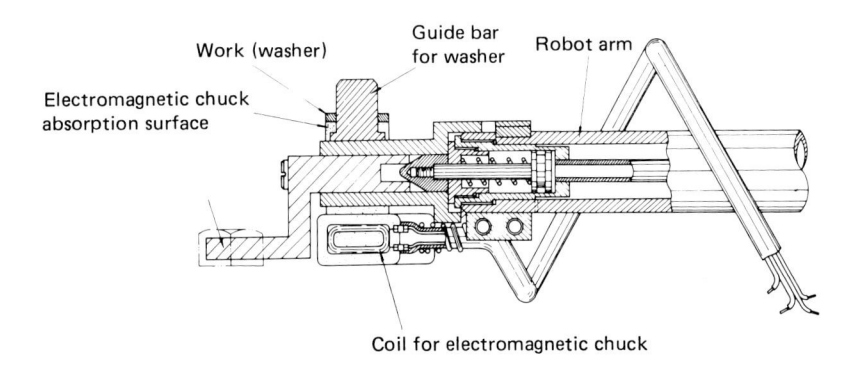

Labels on drawing:
- Work (washer)
- Guide bar for washer
- Robot arm
- Electromagnetic chuck absorption surface
- Coil for electromagnetic chuck

| Air pressure | None | 200 g | 145 (L) × 110 (H) × 45 (W) mm |

FREE TYPE: SNAP

HAND FOR REAR HEAD GRINDING

Daikin Kogyo Co., Ltd.

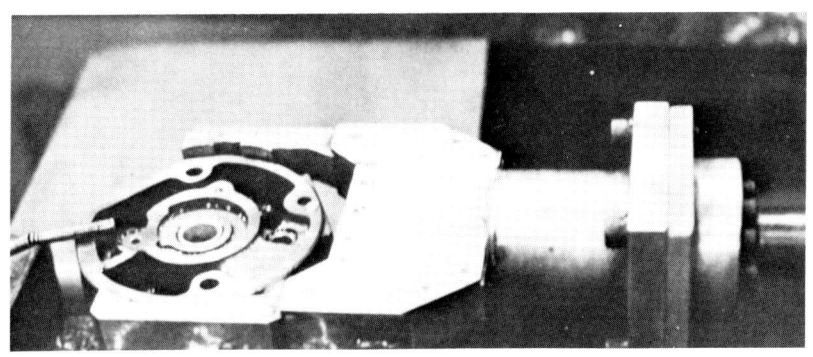

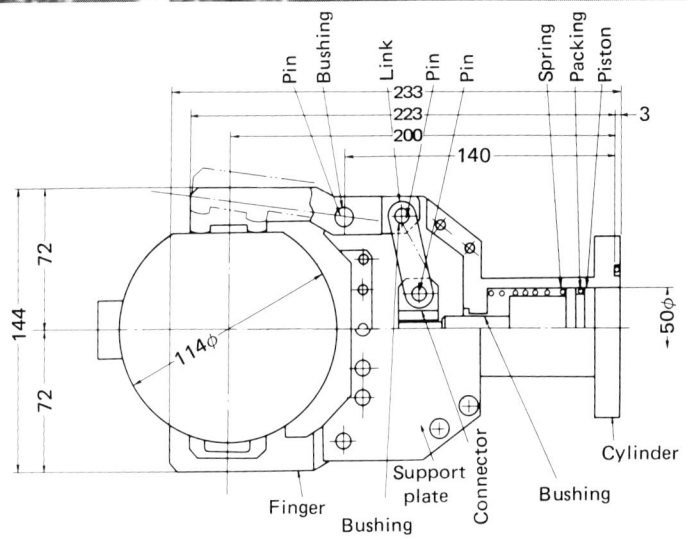

Working Speed				Working Range			
finger	wrist			finger	wrist		
open & shut	up & down	rotation		open & shut	up & down	rotation	
0.5 s		180°/s		114 mm		180°	

Power Source	Sensors	Maximum Load	Size	Weight
Air pressure (4 kg/cm² G)	None	5 kg (total robot)	233 × 144 mm	3.4 kg

NOTE

1. One part of the cylinder need not be supplied with air pressure as the hand can be opened by the compression coil spring of a link mechanism.
2. Compression coil spring design takes into account both the friction between the cylinder packing and bearings and the air exhaust force.
3. Working range is 220 mm back-forth, 70 mm up-down, and 45° right-left; speed is 500 mm/s back-forth and 60°/s right-left.
4. Unvarying sequence control of the robot.
5. On–off control of the finger.
6. Grasping force 75 kg.

THREE-DIRECTION GRIP HAND

Tokyo Keiki Co., Ltd.

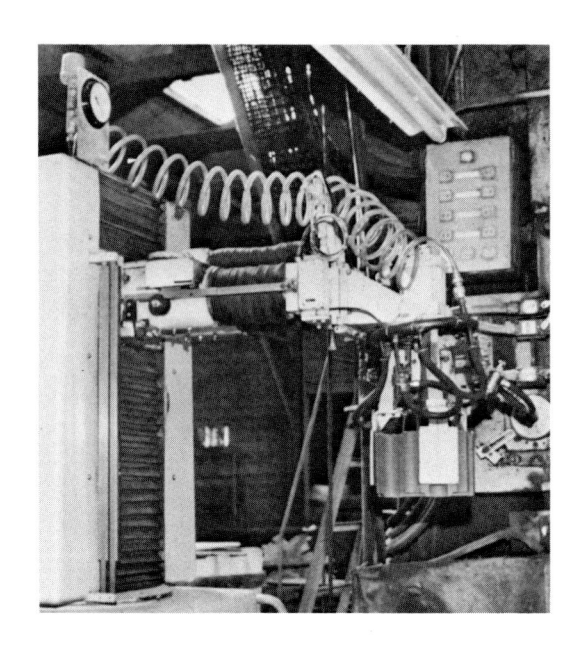

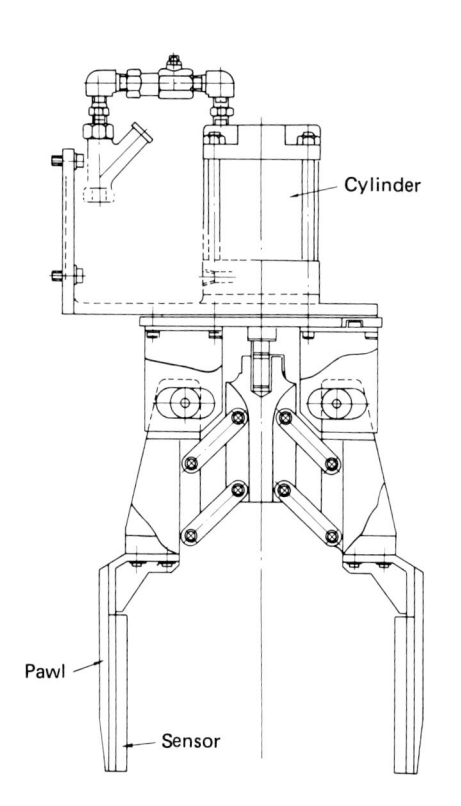

Cylinder

Pawl

Sensor

Working Speed			Working Range		
finger	wrist		finger	wrist	
open & shut	up & down	rotation	open & shut	up & down	rotation
0.5 s/20 mm	Freely settable		20 mm	Freely designable	

Power Source	Sensors	Maximum Load	Size	Weight
Air pressure	Touch sensors (for soft contact)	30 kg	Dependent on work	15 kg

NOTE

1. Two-position control by air pressure.

FREE TYPE: SNAP   **101**

| **PARALLEL GRIP HAND** | **Tokyo Keiki Co., Ltd.** |

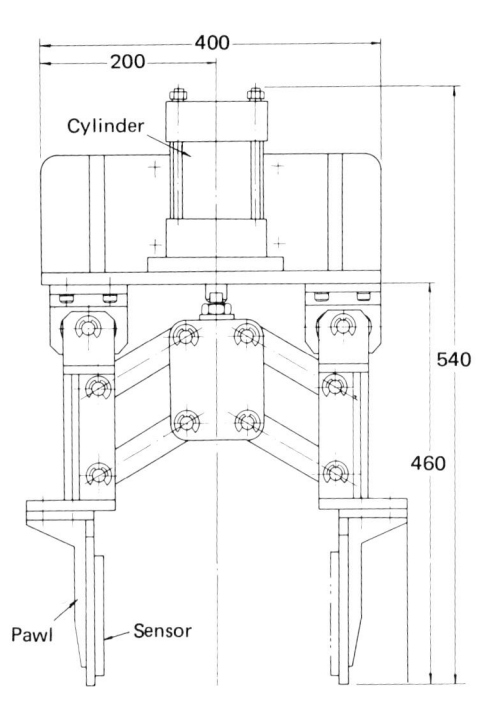

Working Speed		Working Range	
finger	wrist	finger	wrist
open & shut	up & down rotation	open & shut	up & down rotation
0.5 s/40 mm	Freely settable	40 mm	Freely designable

Power Source	Sensors	Maximum Load	Size	Weight
Air pressure	Touch sensors (for soft contact)	45 kg	540 × 400 mm	28 kg

NOTE
1. Two-position control by air pressure.

SINGLE THREE ARMS PLUS JAW TWISTING UNIT (R-4)

Okamura Mfg. Co., Ltd.

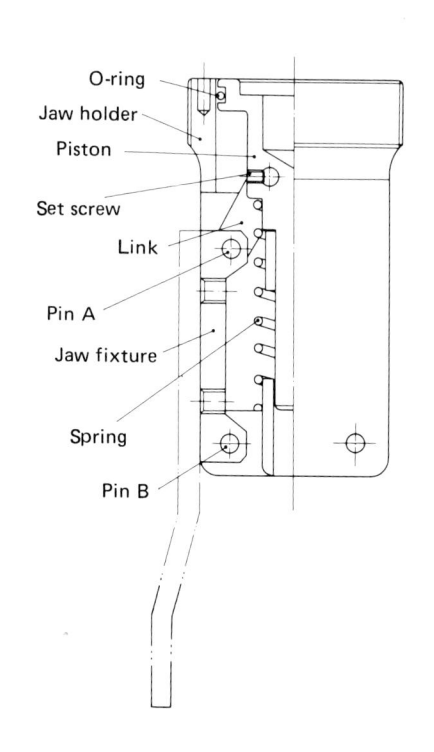

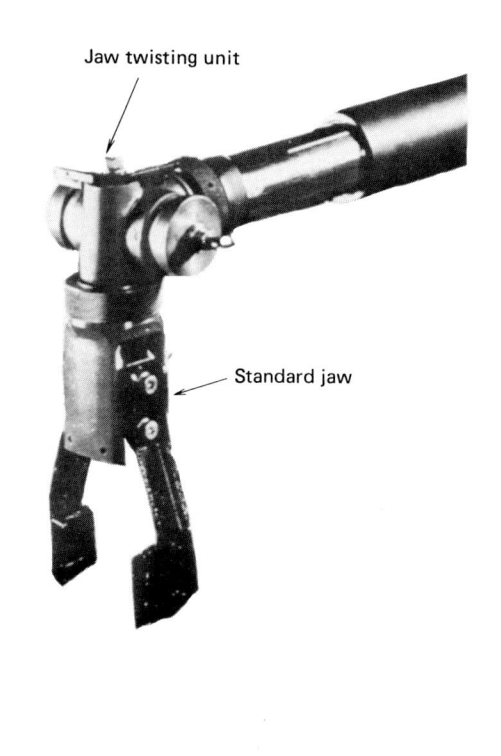

Labels (left diagram): O-ring, Jaw holder, Piston, Set screw, Link, Pin A, Jaw fixture, Spring, Pin B

Labels (right photo): Jaw twisting unit, Standard jaw

Working Speed			Working Range		
finger	wrist		finger	wrist	
open & shut	twist	rotation	open & shut	twist	rotation
0.5 s	180°/s	180°/s	26° (max.)	0–180°	0–180°

Power Source	Sensors	Maximum Load	Size	Weight
Air pressure (4–7 kg/cm² G)	None	5 kg	Dependent on work	15 kg (unit weight)

NOTE

1. Program pin board control.
2. Jaw twisting unit added to the single three arms as an option.
3. Working range of the arm is 130 mm up-down, 500 mm back-forth, and 220° right-left; speed is 150 mm/s, 500 m/s, and 120°/s, respectively.

FREE TYPE: SNAP

CLAMP JAW	Keiaisha Mfg. Co., Ltd.

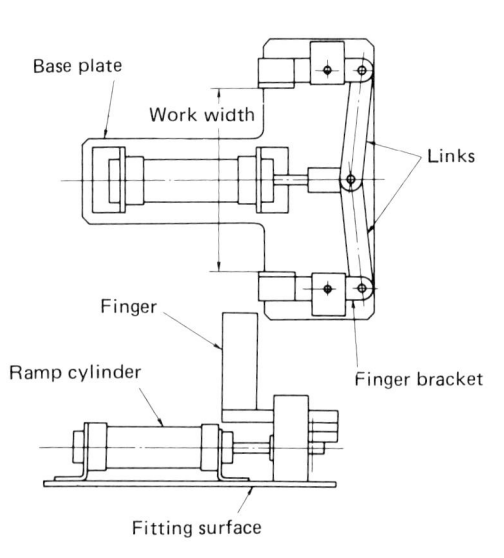

Labels on diagram: Base plate, Work width, Links, Finger, Ramp cylinder, Finger bracket, Fitting surface

Working Speed			Working Range		
finger	wrist		finger	wrist	
open & shut	up & down	rotation	open & shut	up & down	rotation
Free	Freely settable		Free	Freely designable	

Power Source	Sensors	Maximum Load	Size	Weight
Air pressure	None	Dependent on work	Free	5.3 kg

NOTE
1. Cylinder's reciprocal motion is conducted to the links, and the fingers move in an arc to clamp; applied to boxlike work.

CLAMP JAW

Keiaisha Mfg. Co., Ltd.

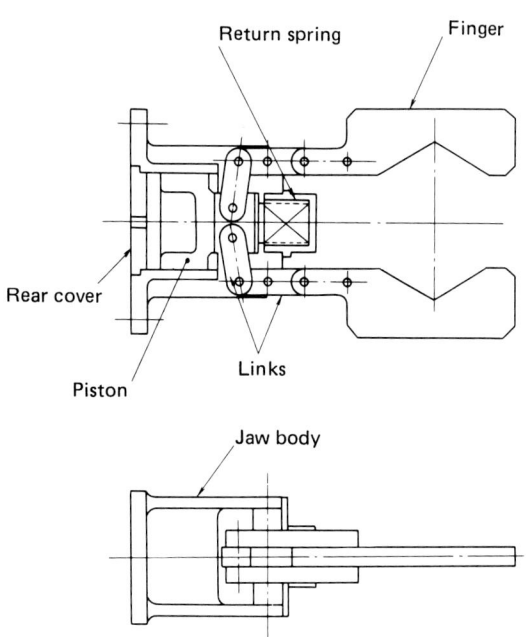

Working Speed			Working Range		
finger	wrist		finger	wrist	
open & shut	up & down	rotation	open & shut	up & down	rotation
Free	Freely settable		Free	Freely designable	

Power Source	Sensors	Maximum Load	Size	Weight
Air pressure	None	10 kg	Freely designable	5.3 kg

NOTE

1. Cylinder's reciprocal motion is conducted to the links, and the fingers move in an arc to clamp; mainly applied to cylindrical work.

FREE TYPE: SNAP

HAND FOR STACKING ALUMINUM INGOTS — Shinko Electric Co., Ltd.

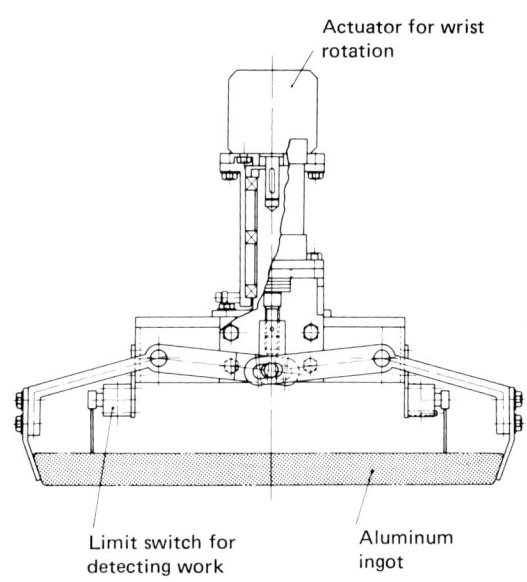

Actuator for wrist rotation

Limit switch for detecting work

Aluminum ingot

Working Speed				Working Range			
finger	wrist			finger	wrist	
open & shut	up & down	rotation		open & shut	up & down	rotation
0.3 s	90°/s			40°	Freely designable	

Power Source	Sensors	Maximum Load	Size	Weight
Air pressure	None	25 kg	Dependent on work	20 kg

NOTE

1. Made for stacking aluminum ingots.
2. Four or five ingots can be grasped at a time.
3. Link-method jaw with an air cylinder.
4. Working range of the arm is 500 mm up-down, 600 mm back-forth, and 60° right-left; speed is 300 mm/s, 500 mm/s, and 90°/s, respectively.

HAND FOR CARRYING COPPER PIPES

Shinko Electric Co., Ltd.

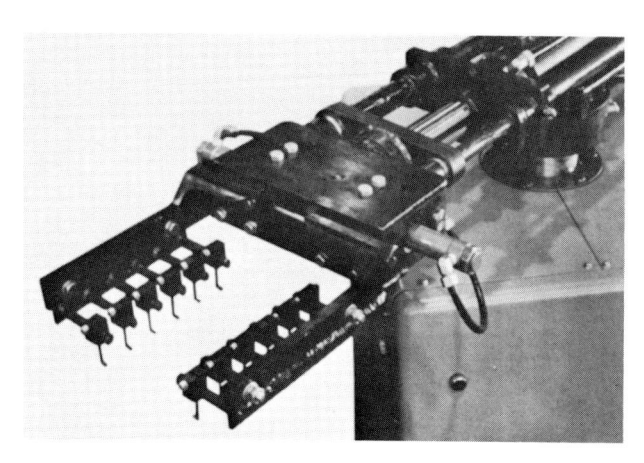

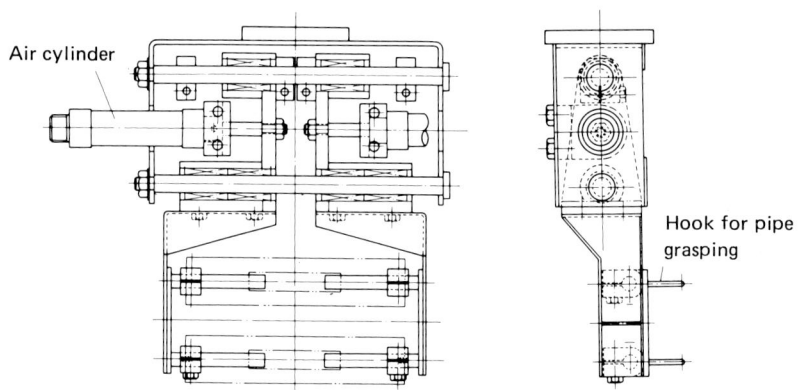

Air cylinder

Hook for pipe grasping

Working Speed			Working Range	
finger	wrist		finger	wrist
open & shut	up & down	rotation	open & shut	up & down rotation
Free	Freely settable		Dependent on work	Freely designable

Power Source	Sensors	Maximum Load	Size	Weight
Air pressure	None	5 kg	Freely designable according to work	10 kg

NOTE

1. Used to carry copper pipes for a bulge work press.
2. Parallel moving-type jaw with two air cylinders.
3. Working range of the arm is 50 mm up-down, 500 mm back-forth, and 90° right-left; speed is 300 mm/s, 1000 mm/s, and 90°/s, respectively.

FREE TYPE: SNAP

KHR−800 W | Star Seiki Co., Ltd.

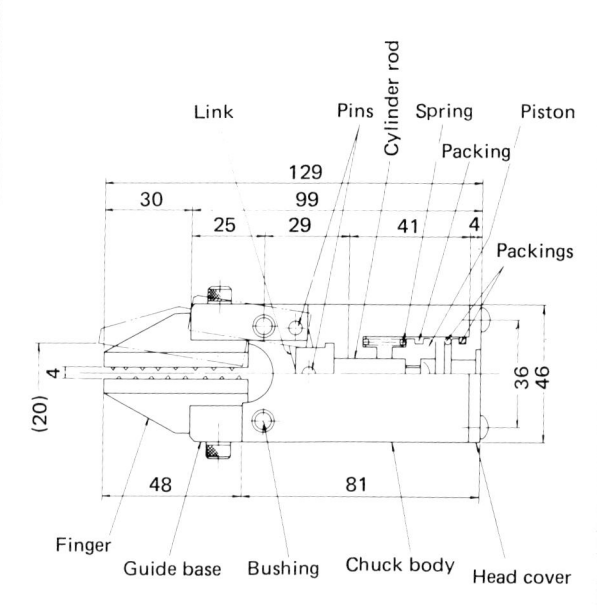

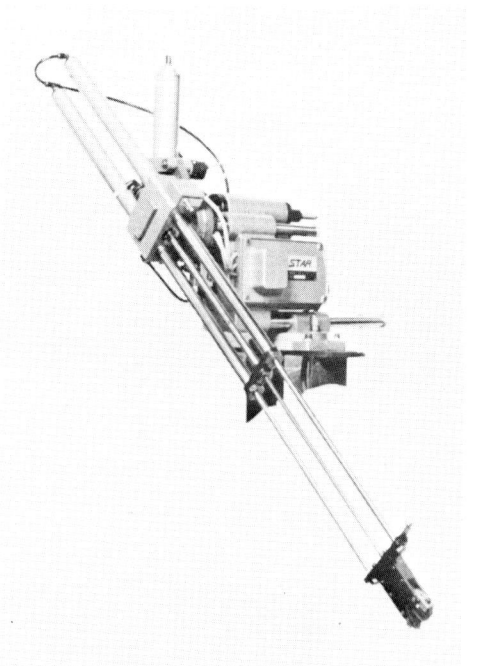

Labels on diagram: Link, Pins, Cylinder rod, Spring, Piston, Packing, Packings, Finger, Guide base, Bushing, Chuck body, Head cover

Dimensions: 30, 129, 25, 99, 29, 41, 4, (20), 4, 36, 46, 48, 81

Working Speed		Working Range	
finger	wrist	finger	wrist
open & shut	up & down rotation	open & shut	up & down rotation
0.2 s	Freely settable	30 mm	Freely designable

Power Source	Sensors	Maximum Load	Size	Weight
Air pressure	Reflex phototube limit switches	2 kg	600 × 250 × 800 mm	51 kg

NOTE

1. Unvarying sequence control.
2. Working range of the arm is 740 mm up-down, 100 mm back-forth, and 0–60° right-left; speed is 600 mm/s, 300 mm/s, and 60°/s, respectively.

HAND FOR TV FRAME

Star Seiki Co., Ltd.

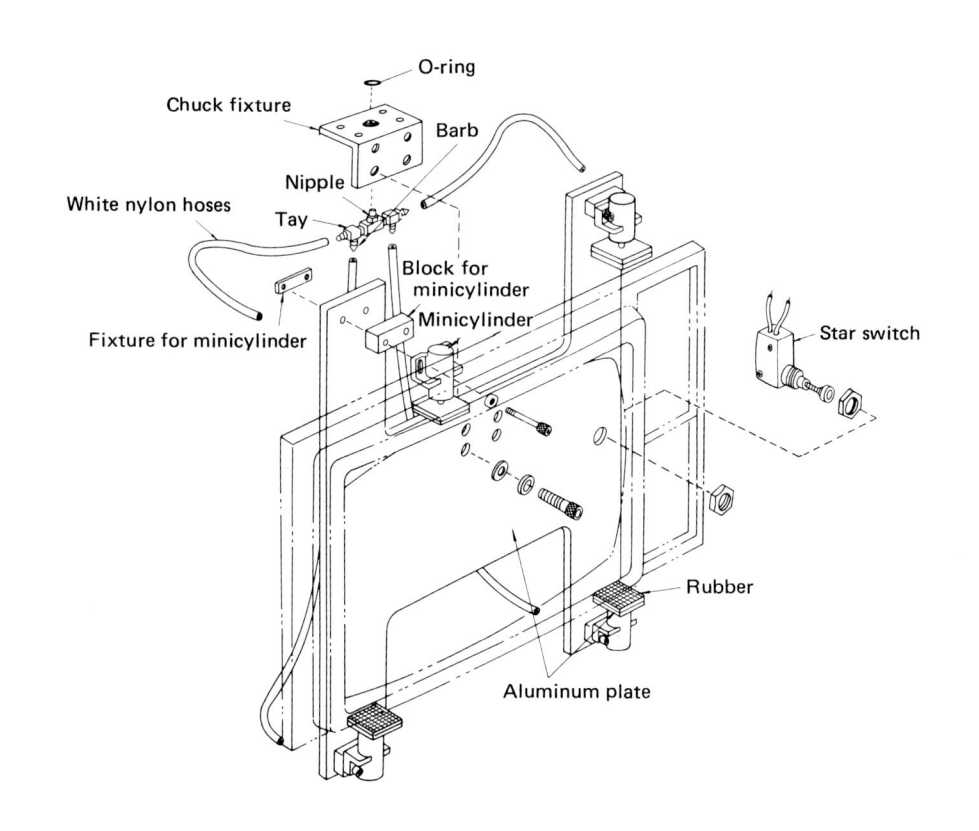

Working Speed			Working Range		
finger	wrist		finger	wrist	
open & shut	up & down	rotation	open & shut	up & down	rotation
0.5 s	1 s/90°		50 mm		90° (fixed)

Power Source	Sensors	Maximum Load	Size	Weight
Air pressure	Reflex phototube limit switches	20 kg	1700 × 2600 × 1850 mm	850 kg

NOTE

1. Force-sparing fundamental parts of the hand are standardized.
2. Excels in general utility for its combination of parts.
3. Low in cost, light, and easy to make.
4. Control by unvarying sequence and mechanical stoppers.
5. Working range of the arm is 1250 mm up-down, 500 mm back-forth, and 1900 mm right-left; speed is 500 mm/s, 300 mm/s, and 300 mm/s, respectively.

FREE TYPE: SNAP

HAND FOR BROWN TUBES

Toshiba Co.

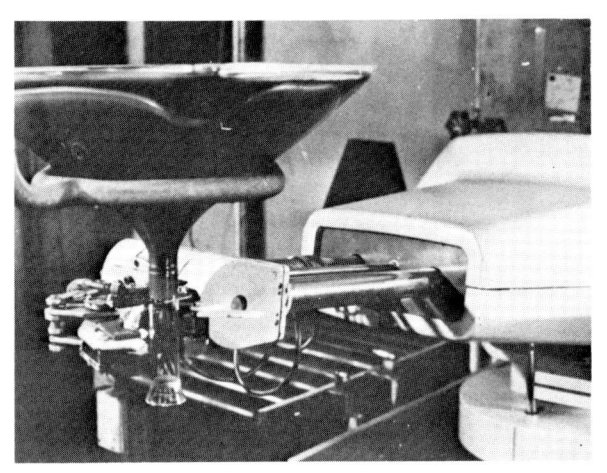

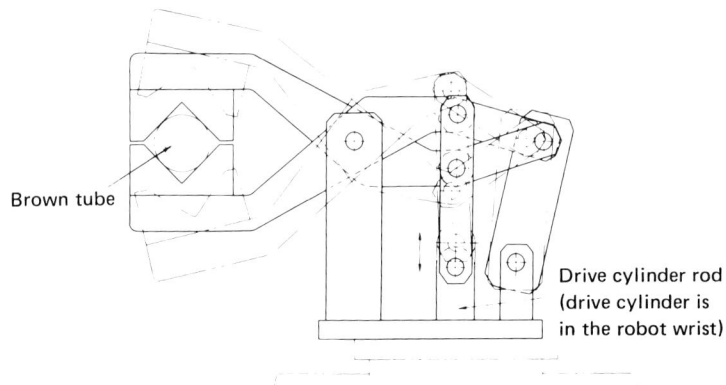

Brown tube

Drive cylinder rod (drive cylinder is in the robot wrist)

Working Speed				Working Range			
finger	wrist			finger	wrist	
open & shut	up & down	rotation		open & shut	up & down	rotation
0.3 s	90°/s	90°/s	30°		220°	180°

Power Source	Sensors	Maximum Load	Size	Weight
Air pressure	None	10 kg		3 kg

NOTE

1. Used for handling TV Brown tubes. Direction of the hand shifts 90° from the direction of the cylinder rod, as the handling cylinder is set in the robot.
2. Pawl of the hand is coated with rubber and can handle glass material such as Brown tubes.
3. Effective in installing a cylinder shifted 90° from the hand.
4. Grasping force 40 kg, cylinder stroke for the hand 30 mm.

GENERAL UTILITY HAND

Toshiba Seiki Co., Ltd.

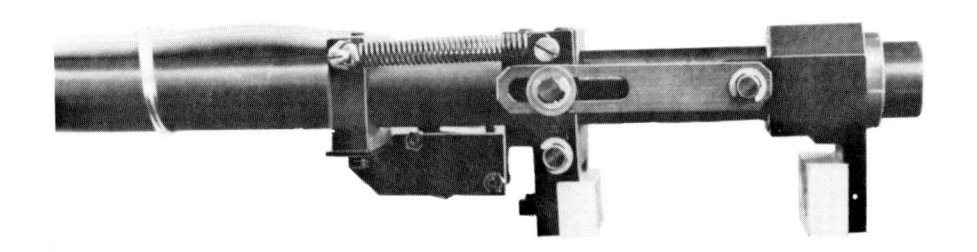

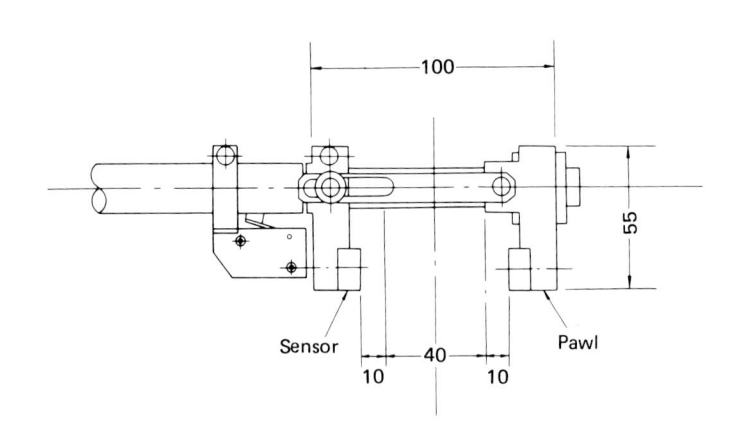

Sensor Pawl
100 55 10 40 10

Working Speed			Working Range		
finger	wrist		finger	wrist	
open & shut	up & down	rotation	open & shut	up & down	rotation
0.5 s	Free		20 mm	Free	

Power Source	Sensors	Maximum Load	Size	Weight
Air pressure	Air sensors	0.5 kg		700 g

NOTE

1. Hand grasps the center of a part; made for general utility.
2. Mainly used for square bars and rods.
3. Working range of the arm is 40 mm up-down, 60–120 mm right-left; speed is 300 mm/s and 180°/s, respectively.

FREE TYPE: SNAP

MECHANICAL HAND WITH VACUUM CHUCK | Toshiba Seiki Co., Ltd.

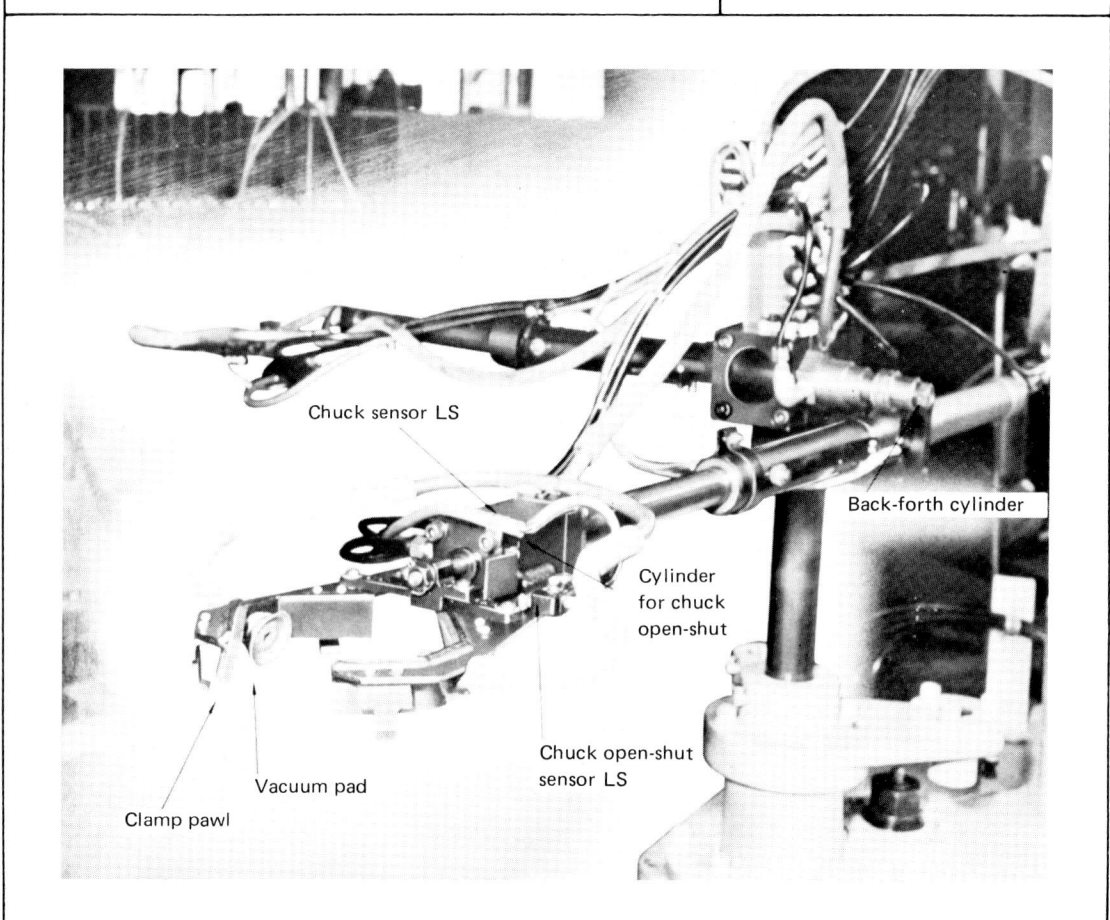

Chuck sensor LS
Back-forth cylinder
Cylinder for chuck open-shut
Chuck open-shut sensor LS
Vacuum pad
Clamp pawl

Working Speed				Working Range			
finger	wrist		finger	wrist		
open & shut	up & down	rotation	open & shut	up & down	rotation	
0.5 s	Free		60 mm	Free		

Power Source	Sensors	Maximum Load	Size	Weight
Air pressure	Air sensors	2 kg		1 kg

NOTE

1. Used to handle conical bodies as in pulling reduction of area work by a press machine.
2. Equipped with a mechanical chuck and a vacuum chuck; most suitable for slippery tapered parts.
3. Working range of the arm is 150 mm back-forth, 90° right-left; speed is 150 mm/s and 180°/s, respectively.

| HAND WITH CHUCKS ON BOTH SIDES | Seiko-Seiki Co., Ltd. |

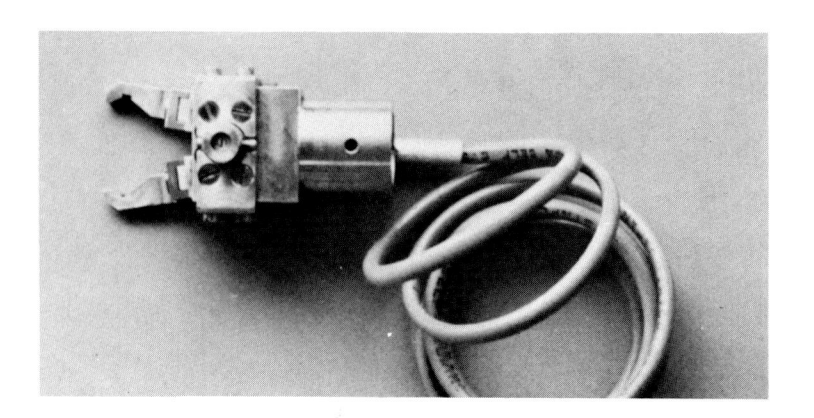

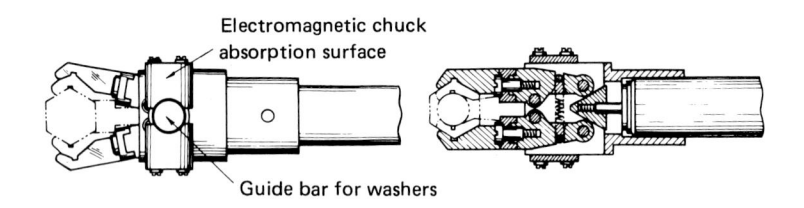

Electromagnetic chuck absorption surface

Guide bar for washers

Working Speed			Working Range		
finger	wrist		finger	wrist	
open & shut	up & down	rotation	open & shut	up & down	rotation
0.5 s	Freely settable		20°	40 mm	Free

Power Source	Sensors	Maximum Load	Size	Weight
Electricity (DC 24 V), air pressure (4 kg/cm² G)	None	100 g (electromagnetic chuck), 150 g (for parallel moving)	90 (L) × 45 (W) × 50 (H) mm	380 g

NOTE

1. Has two functions—grasping and electromagnetic chucking—which are chosen by the arm's inversion. In the grasping phase, bolts and nuts are held; in the electromagnetic phase, washers are held. Material of the parts is selected with care for the electromagnetic chuck.

FREE TYPE: GRIP

SOFT GRIPPER

Tokyo Institute of Technology

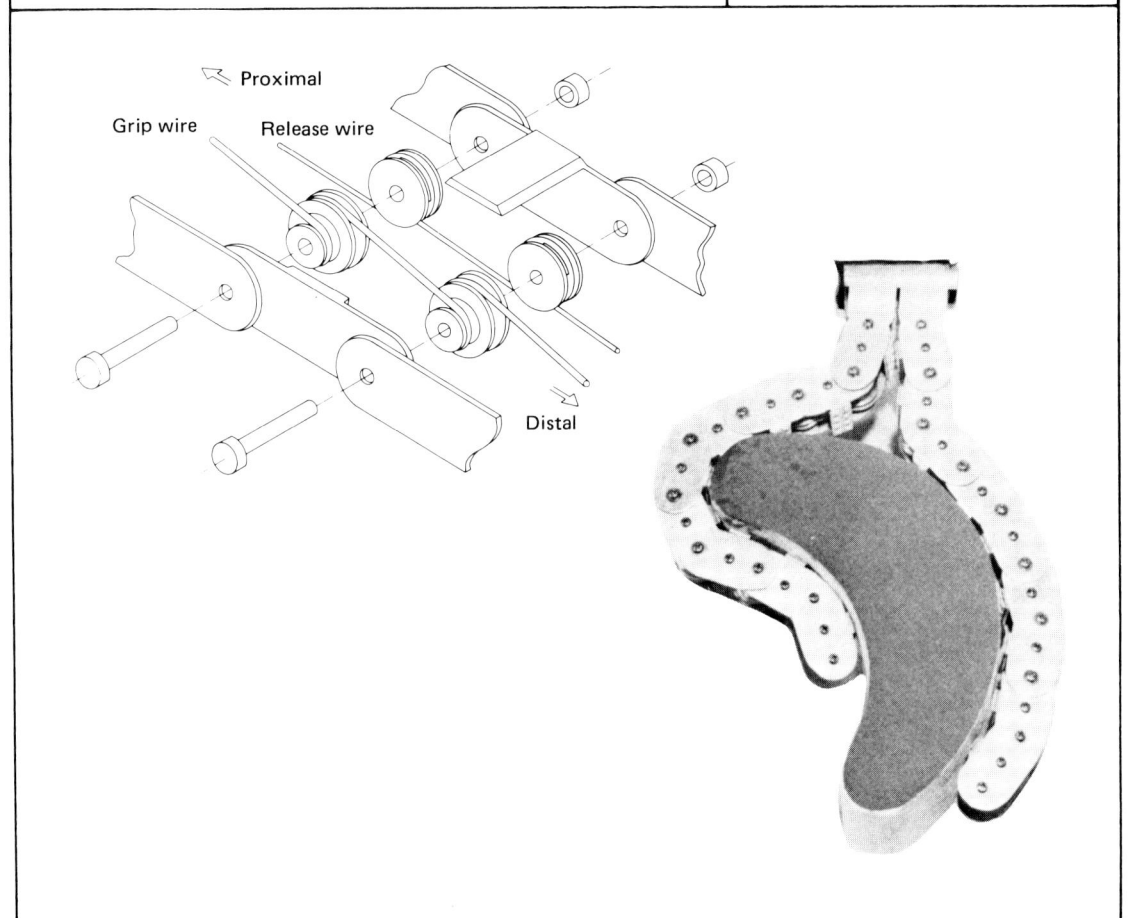

Working Speed			Working Range		
finger	wrist		finger	wrist	
open & shut	up & down	rotation	open & shut	up & down	rotation
0.5 s	Freely settable		180°	Freely designable	

Power Source	Sensors	Maximum Load	Size	Weight
Any source, including air, electricity, and oil pressure	None, but installable	2 kg	300 mm	5 kg

NOTE

1. Fitting for all shapes including concave and convex; easy control.
2. Grasps with arbitrary pressure.
3. Designable to give arbitrary pressure distribution according to the object grasped.
4. Large sized gripper applicable for rehabilitation and/or rescue (under development).

HAND FOR FRONT HEADS

Daikin Kogyo Co., Ltd.

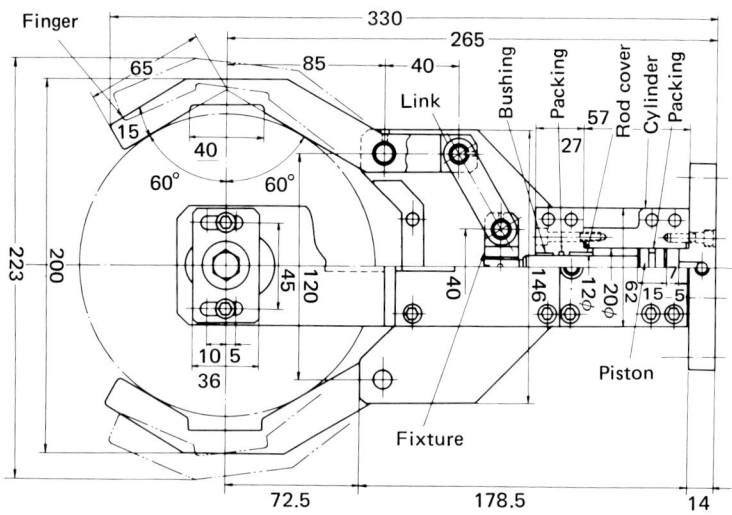

Working Speed				Working Range		
finger	wrist		finger	wrist	
open & shut	swing	rotation	open & shut	up & down	rotation
0.4 s	120°/s	120°/s	200 mm	180°	

Power Source	Sensors	Maximum Load	Size	Weight
Oil pressure (70 kg/cm²)	None	10 kg (total robot)	330 × 223 mm	5.5 kg

NOTE

1. The link mechanism used amplifies the cylinder output, and the work contact against the jig is accomplished by a work press spring.
2. Each bearing of the link mechanism is designed with care for the application of amplified force.
3. Working range of the arm is 300 mm both up-down and back-forth and 180° right-left; speed is 500 mm/s, 700 mm/s, and 110°/s, respectively.
4. Memory recovery servo control in the robot.
5. On–off control in the finger.
6. Grasping force 150 kg.

FREE TYPE: GRIP

OIL PRESSURE HAND FOR CYLINDRICAL STEEL MATERIAL

Marol Company, Ltd.

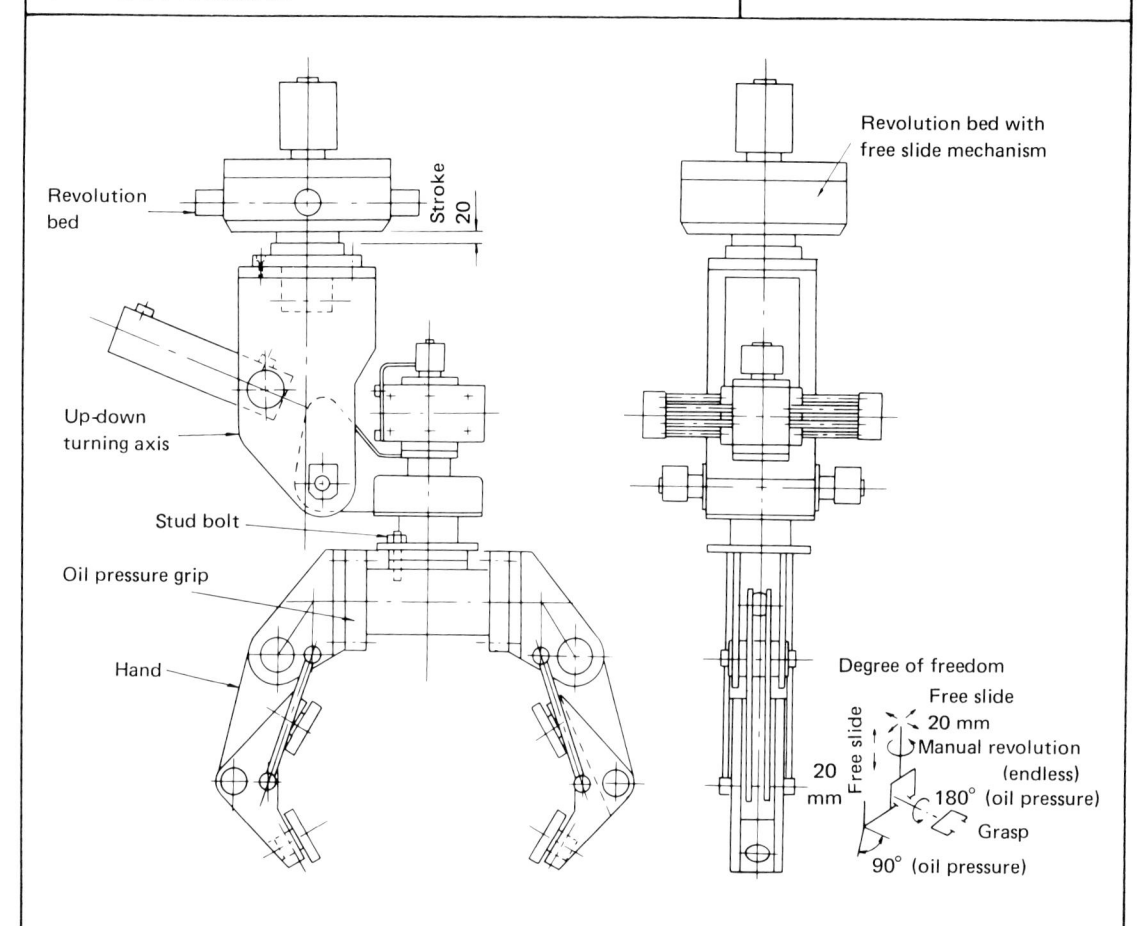

Labels on figures:
- Revolution bed
- Stroke 20
- Up-down turning axis
- Stud bolt
- Oil pressure grip
- Hand
- Revolution bed with free slide mechanism
- Free slide 20 mm
- Degree of freedom
  - Free slide 20 mm
  - Manual revolution (endless)
  - 180° (oil pressure)
  - Grasp
  - 90° (oil pressure)

Working Speed				Working Range			
finger	wrist			finger	wrist	
open & shut	up & down	rotation		open & shut	up & down	rotation
Freely adjustable	5 s/90°	3 s/90°		200–450 mm	90°	180°

Power Source	Sensors	Maximum Load	Size	Weight
Oil pressure	None	350 kg		200 kg

NOTE

1. Hand for carrying cylindrical steel material.
2. Having no sensors, it is equipped with a 20-mm free slide (escape pin) in both the x and y directions.

GRIPPER FOR PULLING OUT IN DIE CASTING

Showa Kuatsuki Co., Ltd.

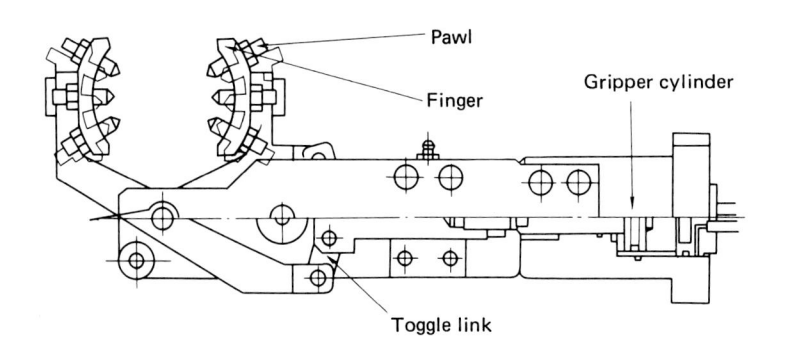

Labels: Pawl, Finger, Gripper cylinder, Toggle link

Working Speed				Working Range			
finger	wrist			finger	wrist	
open & shut	up & down	rotation		open & shut	up & down	rotation
10°/s		90°		5°		90°

Power Source	Sensors	Maximum Load	Size	Weight
Air pressure	None	5 kg	55 mm φ × 28 (W) mm in work grip	600 kg (total robot)

NOTE

1. Attached to robot used to pull out work in die casting; suitable for pulling out objects and setting them in a trimming press.
2. Adjustable pawls can be fitted to the work.
3. Order: sequence control by pin board matrix.
4. Position: mechanical stoppers.
5. Time: timer.
6. Velocity: preset method.

FREE TYPE: GRIP

DOWNWARD MACHINE FINGERS FOR TRANSFERRING

Kyobo

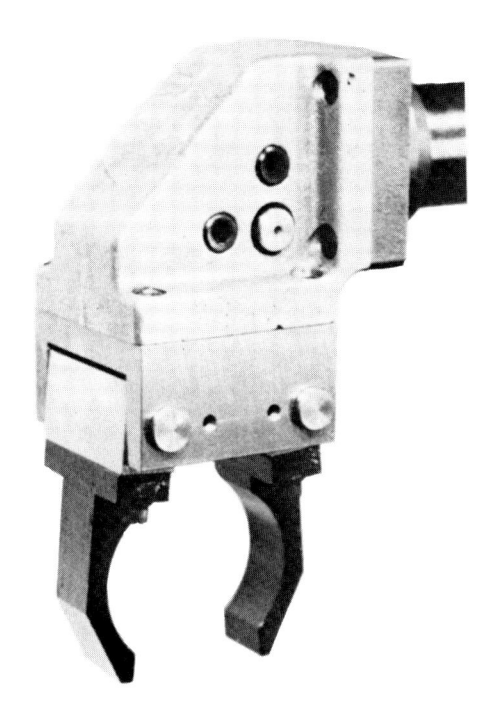

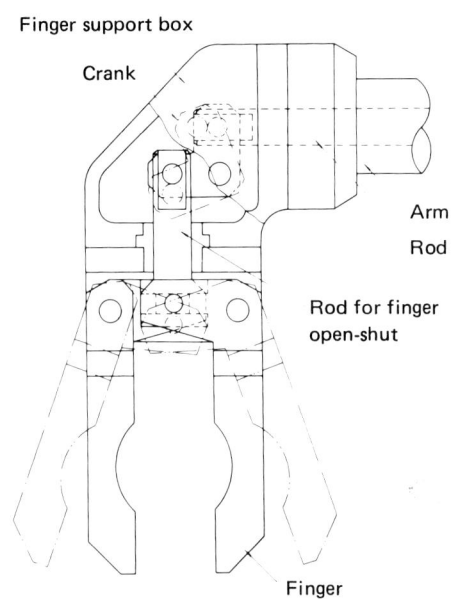

Finger support box
Crank
Arm
Rod
Rod for finger open-shut
Finger

Working Speed			Working Range		
finger	wrist		finger	wrist	
open & shut	up & down	rotation	open & shut	up & down	rotation
1 s		90°/s	45°		90°

Power Source	Sensors	Maximum Load	Size	Weight
Air pressure	None	15 kg	Freely designable	175 kg (total robot)

NOTE

1. Hand for transferring usual machine work, assemblies, and so on.
2. Has the most general mechanism with fingers opened and shut by a bell crank according to back-forth motion of the air cylinder.
3. Working range of the arm is 10 mm up-down, 200 mm back-forth, and 90° right-left; speed is 100 mm/s, 200 mm/s, and 90°/2 s, respectively.

MACHINE FINGERS FOR TRANSFERRING — Kyobo

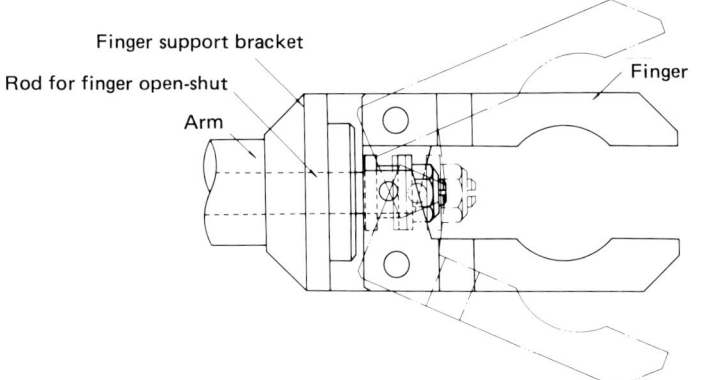

Finger support bracket
Rod for finger open-shut
Arm
Finger

Working Speed			Working Range		
finger	wrist		finger	wrist	
open & shut	up & down	rotation	open & shut	up & down	rotation
1 s		90°/s	45°		90°

Power Source	Sensors	Maximum Load	Size	Weight
Air pressure	None	15 kg	Freely designable	175 kg (total robot)

NOTE

1. Hand for transferring usual machine work, assemblies, and so on.
2. Fingers opened and shut by a bell crank according to back-forth motion of the air cylinder; the finger can have four (up-down and right-left) directions.
3. Working range of the arm is 100 mm up-down, 200 mm back-forth, and 90° right-left; speed is 100 mm/s, 200 mm/s, and 90°/2 s, respectively.

FREE TYPE: GRIP

HIGH-SPEED MACHINE FINGERS FOR TRANSFERRING

Kyobo

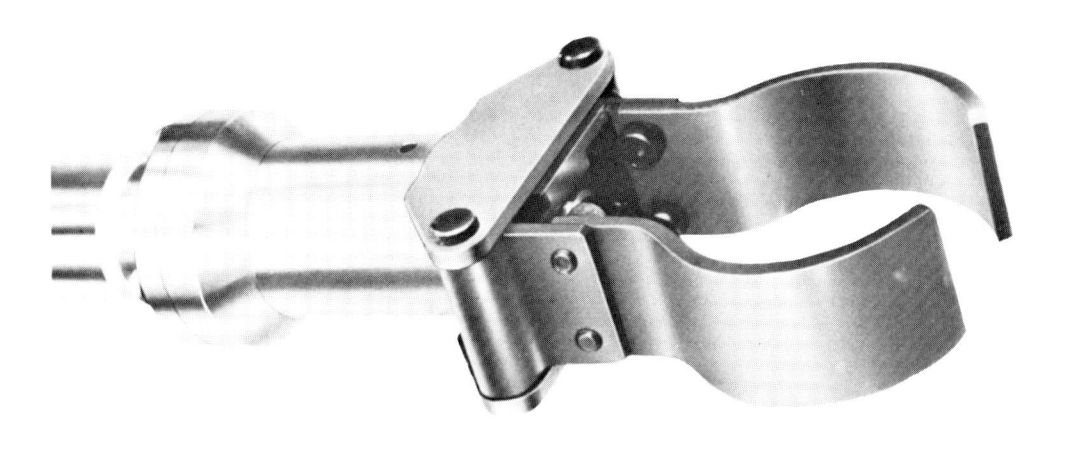

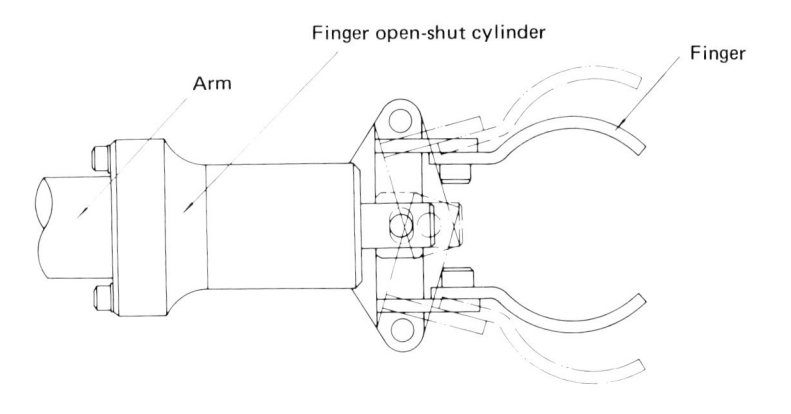

Finger open-shut cylinder
Arm
Finger

Working Speed				Working Range		
finger	wrist		finger	wrist	
open & shut	up & down	rotation	open & shut	up & down	rotation
0.5 s	Free		45°	Freely designable	

Power Source	Sensors	Maximum Load	Size	Weight
Air pressure	None	2 kg	Freely designable	60 kg (total robot)

NOTE

1. Hand for transferring relatively light parts such as press parts and spot welding parts.
2. Clamping is done by a spring and unclamping by an air cylinder.
3. Working range of the arm is 50 mm up-down and 90° right-left; speed is 50 mm/0.5 s, and 90°/s, respectively.

EMBRACER 3, 4

FIT Co.

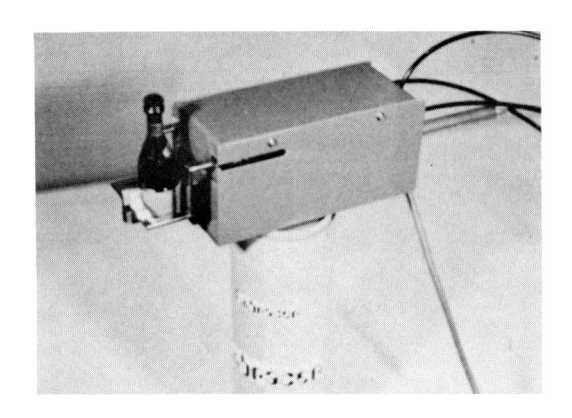

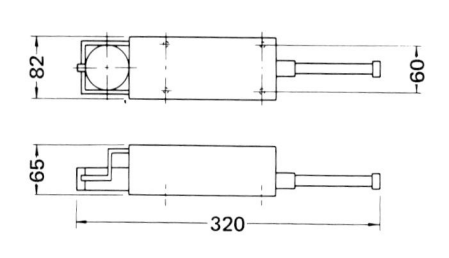

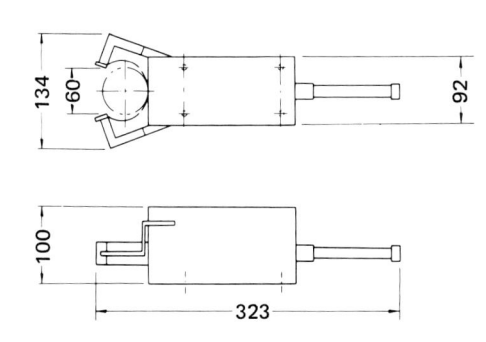

Working Speed			Working Range		
finger	wrist		finger	wrist	
open & shut	up & down	rotation	open & shut	up & down	rotation
Less than 0.1 s			50–500 mm	Freely designable	
Power Source	**Sensors**		**Maximum Load**	**Size**	**Weight**
Air pressure	Sensor for misgrips		3 kg	150 (W) × 150 (H) × 600 (L) mm	2 kg

NOTE

1. For handling fragile work. It is not necessary to change the embracer for work of various sizes.
2. Can grasp softly and tightly because there is no force concentration. Applicable to sintering alloys, raw ceramics, plastics, springs, foods, etc.
3. Flexible soft fingers (without an adjustment).
4. Two finger types: shape 1 with closed tip and shape 2 with open tip.
5. Can grasp work with 70-mm difference in diameter.

FREE TYPE: GRIP

ARTIFICIAL HAND USING WINDTHROW EFFECT

Ministry of Labor Research on Industry and Safety

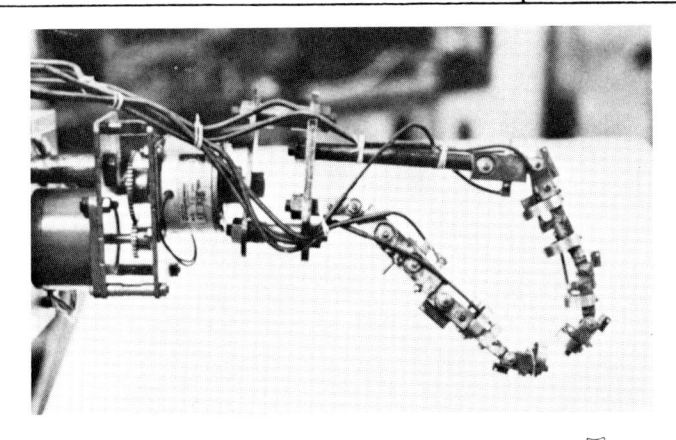

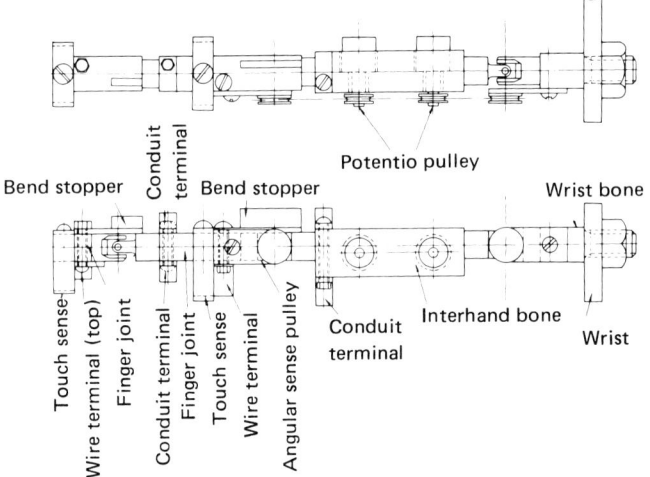

Working Speed			Working Range		
finger	wrist		finger	wrist	
open & shut	up & down	rotation	open & shut	up & down	rotation
2 s	120°/s (max.)		Like human fingers	360°	

Power Source	Sensors	Maximum Load	Size	Weight
Electricity + viscous fluid	6 strain gauges for angle, 3 potentiometers for pressure	About 5 kg (estimated), 10-kg grasping force	Fingers are 1.5 times larger than human ones	500 g (fingers)

NOTE

1. Control method:
   - Arm: induction motors + clutch break system (revolution)
     DC motors + clutch break system (up-down)
     pulse motors (back-forth)
   - Wrist: pulse motor + break system (rotation)
   - Finger: 2 windthrow clutches (for 6 degrees of freedom) voltage control
2. Microcomputer control of the total system. The windthrow clutch has so-called internal electricity fluid, whose viscosity can be controlled by the external applied field. Pressure and angle control are freely performed in spite of the comparatively small size.

THIN-TYPE VACUUM CHUCK

Toshiba Seiki Co., Ltd.

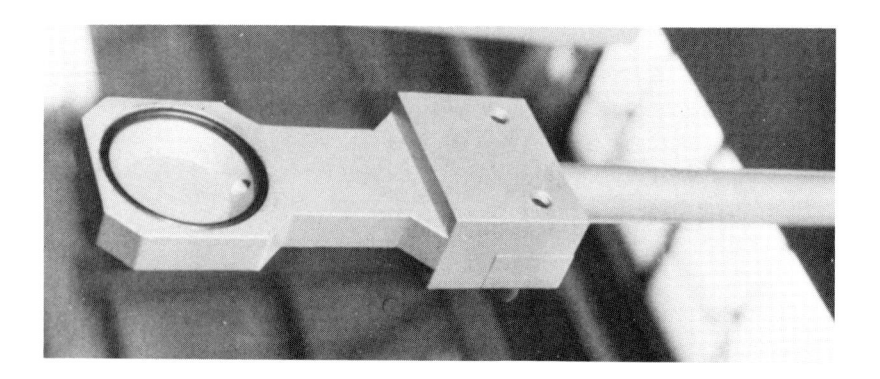

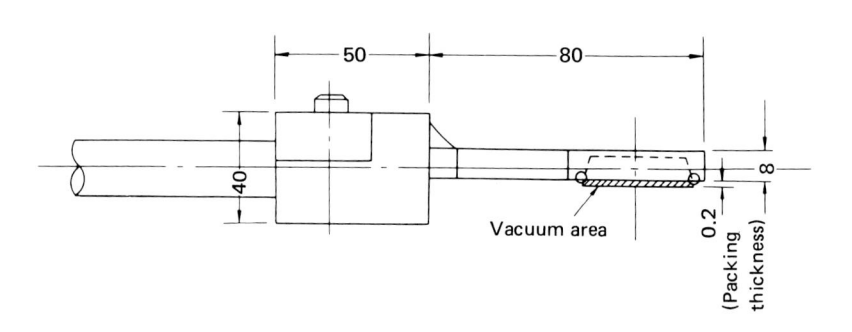

Working Speed			Working Range		
finger	wrist		finger	wrist	
open & shut	up & down	rotation	open & shut	up & down	rotation
0.2 s	Freely settable			Freely designable	

Power Source	Sensors	Maximum Load	Size	Weight
Air pressure (vacuum)	Air sensor	0.1 kg		150 g

NOTE
1. Hand for supplying plate materials and pulling them out.
2. Characteristic thin-chuck thickness of 8 mm.
3. Working range of the arm is 60 mm up-down and 60–120 mm right-left; speed is 300 mm/s and 180°/s, respectively.

FREE TYPE: SUCTION

SUCTION CHUCK FOR THIN PLATES — Toshiba Seiki Co., Ltd.

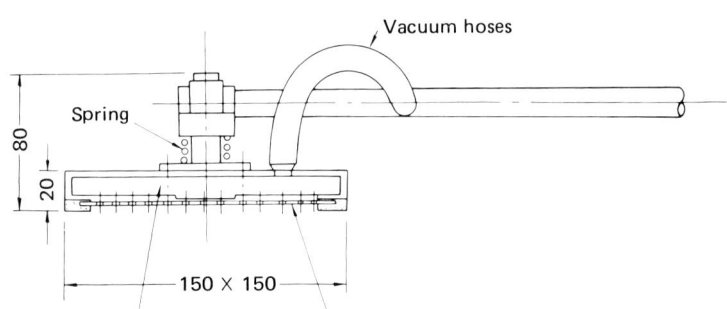

Vacuum hoses
Spring
80
20
150 × 150
Intermediate plate — Iron plate with holes (wire gauze)

Working Speed			Working Range		
finger	wrist		finger	wrist	
open & shut	up & down	rotation	open & shut	up & down	rotation
0.5 s	Freely settable			Freely designable	

Power Source	Sensors	Maximum Load	Size	Weight
Air pressure (vacuum)	Air sensor	0.5 kg		300 g

NOTE
1. For handling soft plate materials that allow no transformation in the vacuum.
2. Best fit for plate materials (thin aluminum, stainless plate, and so on).
3. Working speed of the arm is 300 mm/s up-down and 180°/s right-left; range is 50 mm and 60–120°, respectively.

AUTOMATIC TRACKING VACUUM CHUCK

Toshiba Seiki Co., Ltd.

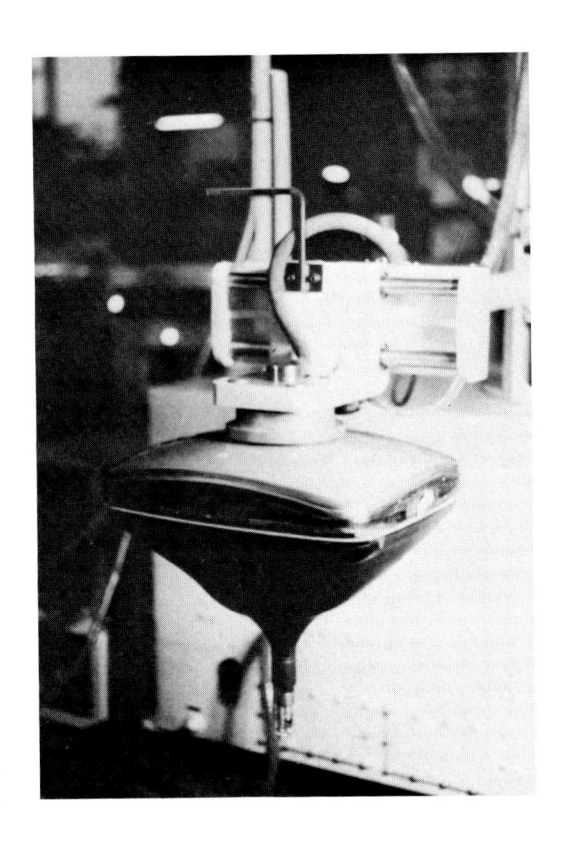

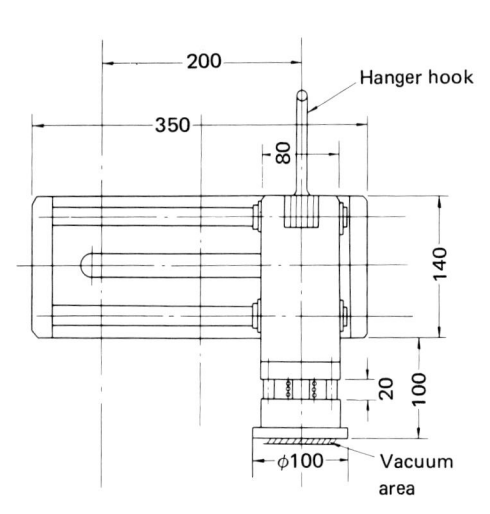

Dimensions: 200, 350, 80, 140, 20, 100, φ100 — Hanger hook, Vacuum area

Working Speed			Working Range		
finger	wrist		finger	wrist	
open & shut	up & down	rotation	open & shut	up & down	rotation
1 s	Freely designable			Freely designable	

Power Source	Sensors	Maximum Load	Size	Weight
Air pressure (vacuum)	Air sensor	15 kg		7 kg

NOTE

1. Equipped with an automatic tracking device, a clamping one and a reset one.
2. Best for supplying and removing objects from a conveyor system.
3. Working range of the arm is 600 mm up-down and 180° right-left; speed is 400 mm/s and 60°/s, respectively.

FREE TYPE: SUCTION

VACUUM HAND (EJECTOR TYPE)	Seilor Pen Co., Ltd.

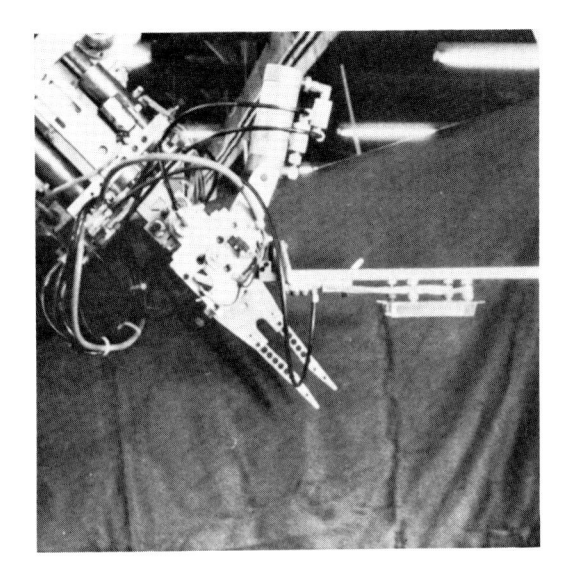

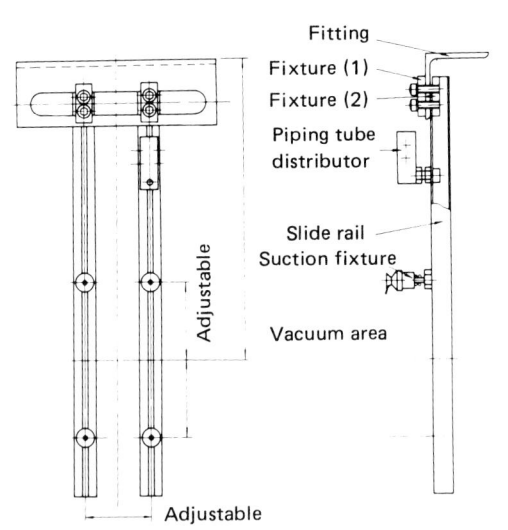

Figure labels: Fitting, Fixture (1), Fixture (2), Piping tube distributor, Slide rail, Suction fixture, Vacuum area, Adjustable

Working Speed			Working Range		
finger	wrist		finger	wrist	
open & shut	up & down	rotation	open & shut	up & down	rotation
0.2 s	90°/s		About 30 mm	90° around the x axis 45° around the y axis	

Power Source	Sensors	Maximum Load	Size	Weight
Air pressure (5.5–6 kg/cm^2)	Limit switches, photo-electric switches	About 1 kg		160 kg (total robot)

NOTE

1. Designed for pulling out plastic emergency mold goods; easy to change work and hands for different work.
2. Has many absorption disks and one-touch exchange.

VACUUM HAND (BLOWER TYPE)

Seilor Pen Co., Ltd.

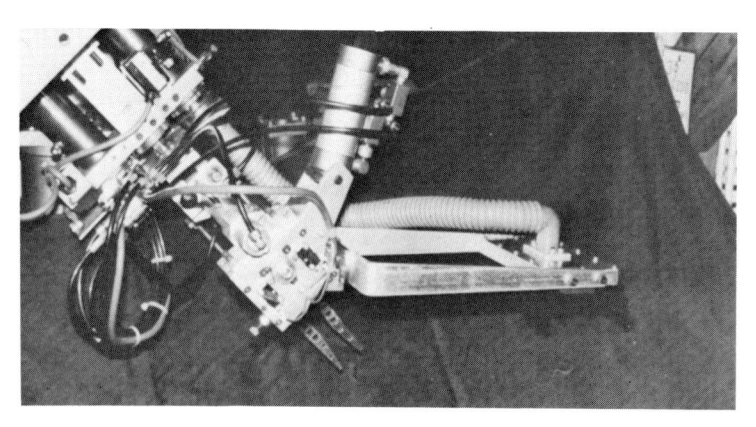

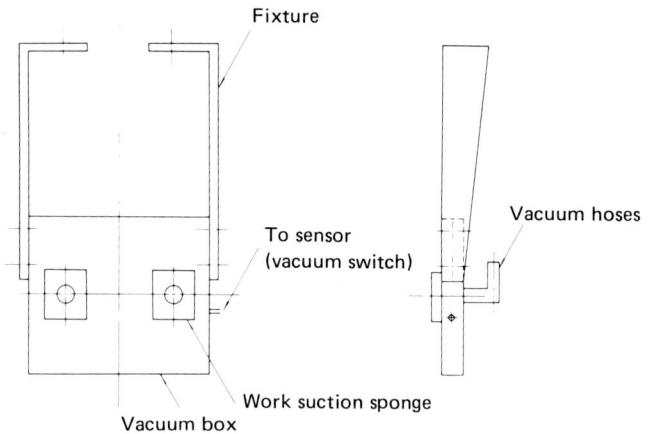

Fixture

To sensor (vacuum switch)

Vacuum hoses

Work suction sponge

Vacuum box

Working Speed				Working Range			
finger	wrist			finger	wrist	
open & shut	up & down	rotation		open & shut	up & down	rotation
0.2 s (absorption time)	90°/s				90° around x axis 45° around y axis	

Power Source	Sensors	Maximum Load	Size	Weight
Air pressure (5.5–6 kg/cm² G)	Vacuum switches	1 kg	1700 (H) × 400 (W) × 600 (L) mm (total robot)	160 kg (total robot)

NOTE

1. Designed for pulling out plastic emergency mold goods; easy to change work and hands for different work (for example, exchange of metal molds).
2. As the airflow is high, suction is possible in spite of the surface of the work with more or less convex and concave parts.
3. Vacuum is about 500 mm Hg.

FREE TYPE: SUCTION

VACUUM HAND

Kawasaki Heavy Industries, Ltd.

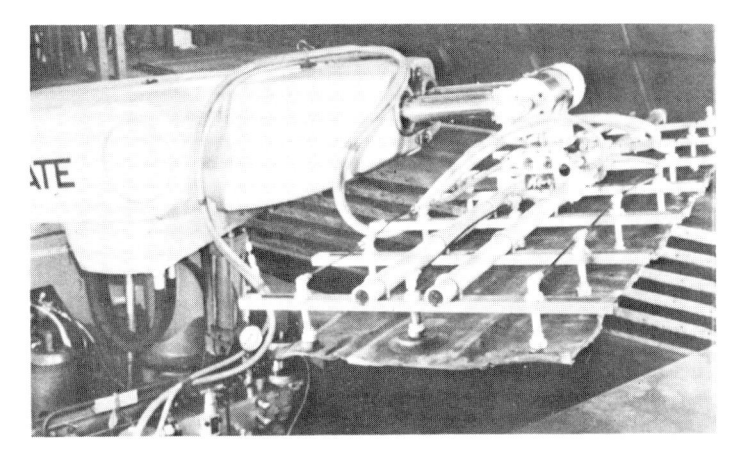

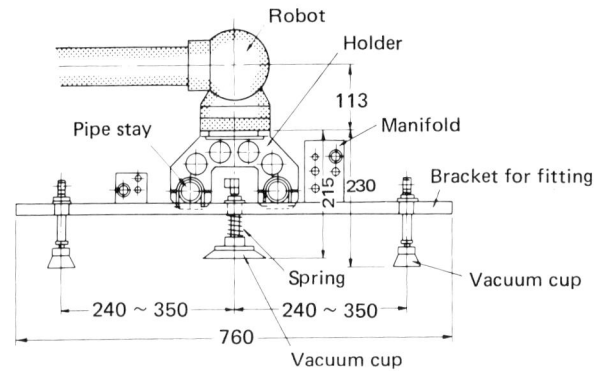

Working Speed			Working Range		
finger	wrist		finger	wrist	
open & shut	up & down	rotation	open & shut	up & down	rotation
	Freely settable			Freely designable	

Power Source	Sensors	Maximum Load	Size	Weight
Air pressure (vacuum pump or venturi)	None	15 kg	760 × 760 mm	10 kg

NOTE

1. For handling soft or hard work such as sheet (steel plates, tire treads, glass plates, and so on).
2. Can add supporting pipes and enlarge vacuum cups according to the size of the work.
3. To improve release of work from the cups, blowing is off when vacuum stops.
4. Power source is a vacuum pump or compressed air in factories (venturi method).
5. Soft materials such as tire treads are protected from damage by adjusting the pressure.

VACUUM SUCTION-TYPE HAND

Yasukawa Electric Mfg. Co., Ltd.

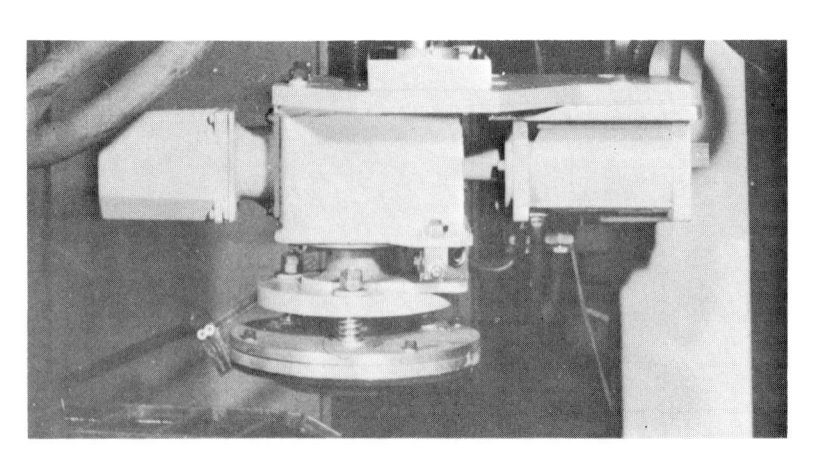

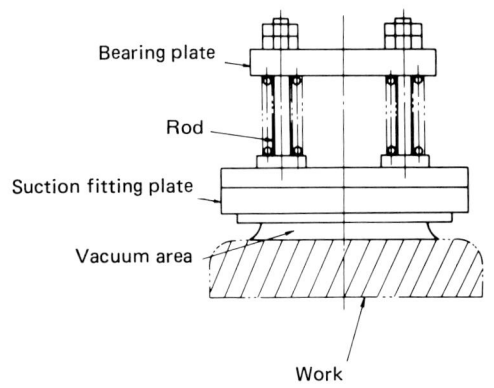

Bearing plate
Rod
Suction fitting plate
Vacuum area
Work

Working Speed			Working Range		
finger	wrist		finger	wrist	
open & shut	up & down	rotation	open & shut	up & down	rotation
0.5 s	Freely designable			Freely designable	

Power Source	Sensors	Maximum Load	Size	Weight
Air pressure (vacuum pump)	None	About 5 kg	About 200 × 200 mm (work)	About 7 kg

NOTE

1. On-off control by timer and pressure switch.

FREE TYPE: SUCTION

HANG ARM I	FIT Co.

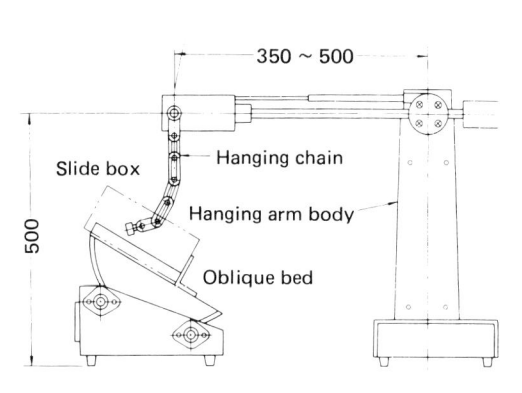

Figure labels: 350 ~ 500; 500; Slide box; Hanging chain; Hanging arm body; Oblique bed

Working Speed			Working Range		
finger	wrist		finger	wrist	
open & shut	up & down	rotation	open & shut	up & down	rotation
Below 0.1 s					

Power Source	Sensors	Maximum Load	Size	Weight
Air or electricity	None	1 kg	300 (L) × 300 (W) × 500 (H) mm	15 kg

NOTE

1. Uses flexibility of a chain. In rotation, the chain is fixed so that it does not swing; the work is set at a given position. However, many objects in the box can be handled as the chain bends.
2. Pulls out one by one when parts are loosely introduced to the production line; best suited for part supply for an automatic machine.
3. Pulls out one by one from loose materials.
4. Gripping by vacuum or electromagnetic absorption.
5. Working range is 300 mm up-down and 600 mm right-left; speed is 4 s/cycle.

HANG ARM II	FIT Co.

Working Speed			Working Range		
finger	wrist		finger	wrist	
open & shut	up & down	rotation	open & shut	up & down	rotation
Below 0.1 s (for gripping)					

Power Source	Sensors	Maximum Load	Size	Weight
Air or electricity	None	1 kg	400 (L) × 2700 (W) × 1800 (H) mm	50 kg

NOTE

1. Pulls out work in order from above and stacks in order; the handling force is the force necessary for only one object. The arm can both pull out and stack by changing the sequence.
2. Used for stacking on a buffer stock and pulling out from the stack in the middle of a line.
3. Maximum thickness of a stacked object is 700 mm.
4. In stacking, the work is not pressed for loose contact.
5. Used for stacking and pulling out stacked work.
6. Gripping by vacuum or electromagnetic force.
7. Working range of the arm is 700 mm up-down and 2.5 m right-left; speed is less than 14 s/cycle.

FREE TYPE: SUCTION

DOUBLE VENTURI-TYPE JAW | Keiaisha Mfg. Co., Ltd.

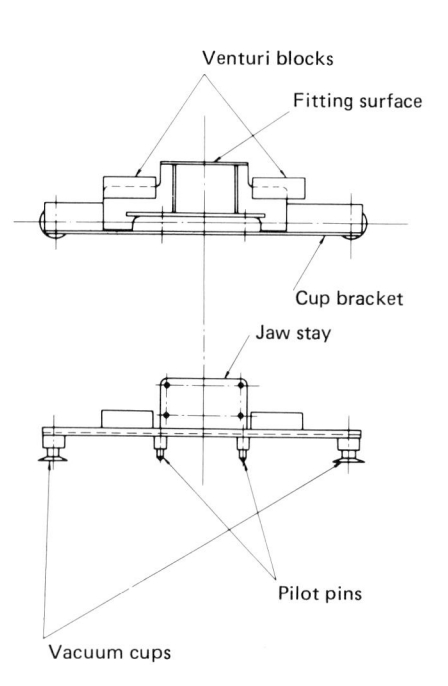

Venturi blocks
Fitting surface
Cup bracket
Jaw stay
Pilot pins
Vacuum cups

Working Speed				Working Range			
finger	wrist			finger	wrist		
open & shut	up & down		rotation	open & shut	up & down		rotation
	Freely designable				Freely designable		

Power Source	Sensors	Maximum Load	Size	Weight
Air pressure	None	7 kg	Freely designable	

NOTE
1. Used for conveying sheet materials which are slippery because of their heavy weight (about 7 kg) and oil coating. They are prevented from slipping by the use of locating pilot pins. Cup brackets are fixed to the jaw stay by two bolts and the brackets can be exchanged for others to handle other types of work.

DOUBLE VENTURI-TYPE JAW

Keiaisha Mfg. Co., Ltd.

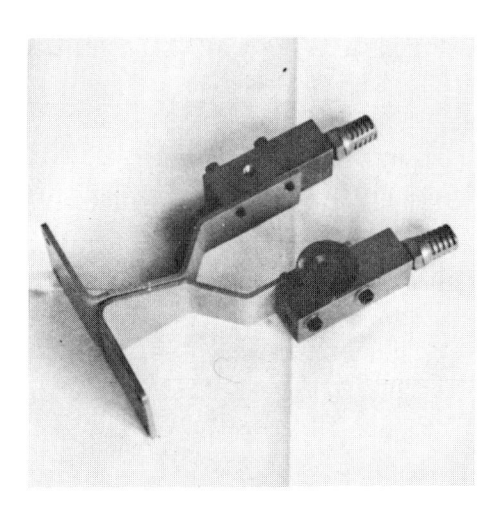

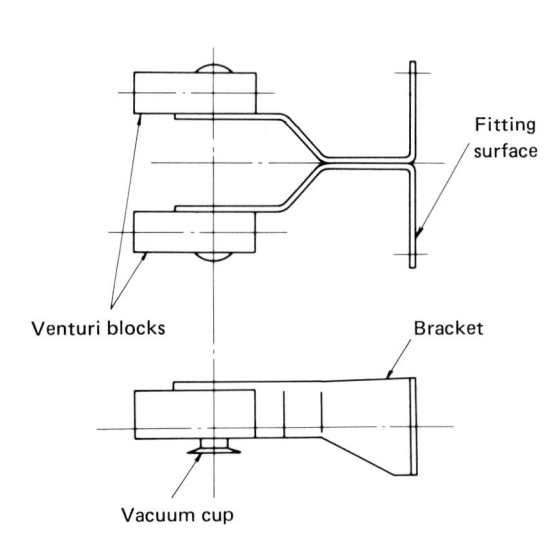

Fitting surface

Venturi blocks

Bracket

Vacuum cup

Working Speed		Working Range	
finger	wrist	finger	wrist
open & shut	up & down rotation	open & shut	up & down rotation
	Freely designable		Freely designable

Power Source	Sensors	Maximum Load	Size	Weight
Air pressure	None	Dependent on work	Freely designable	2.5 kg

NOTE
1. Jaw is mainly used for conveying sheet materials, and two vacuum cups are attached for heavy work.
2. Venturi block uses negative pressure method. Compared with vacuum pump method, suction can be accomplished with one source of air pressure, and the device is light and small. Even if one of the two cups is released, the suction ability of the remaining one does not change.

FREE TYPE: SUCTION

HAND FOR CONVEYING BROWN TUBES

Shinko Electric Co., Ltd.

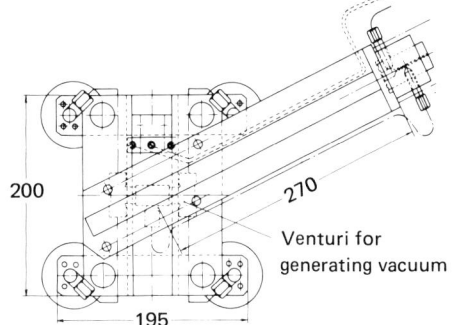

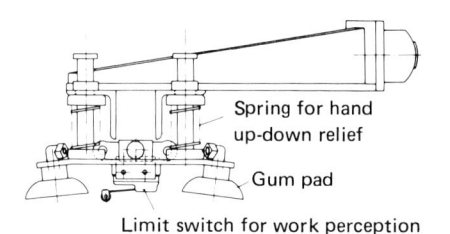

Venturi for generating vacuum

Spring for hand up-down relief

Gum pad

Limit switch for work perception

Working Speed			Working Range		
finger	wrist		finger	wrist	
open & shut	up & down	rotation	open & shut	up & down	rotation
	Freely settable			Freely designable	

Power Source	Sensors	Maximum Load	Size	Weight
Air pressure	Limit switch	10 kg	195 × 200 mm	10 kg

NOTE

1. For conveying TV Brown tubes.
2. Venturi-type vacuum pads with vinyl covers prevent the glass from being marked by gum pads.
3. Working range of the arm is 150 mm up-down, 500 mm back-forth, and 180° right-left; speed is 300 mm/s, 500 mm/s, and 90°/s, respectively.

SUCTION HAND WITH JAW

Shinko Electric Co., Ltd.

Limit switch for dish detecting

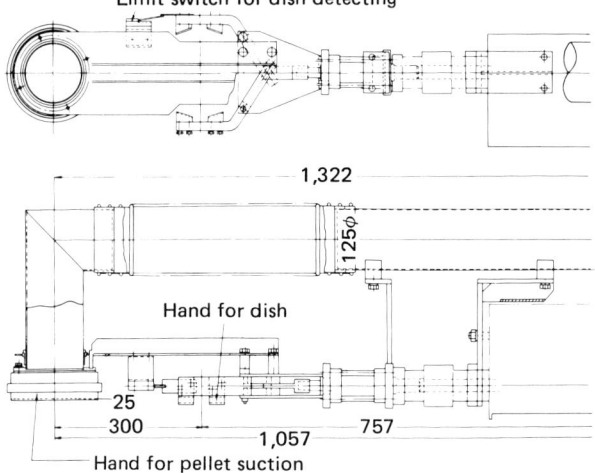

1,322
125φ
Hand for dish
25
300
1,057
757
Hand for pellet suction

Working Speed			Working Range		
finger	wrist		finger	wrist	
open & shut	up & down	rotation	open & shut	up & down	rotation
0.3 s	Free			Freely designable	

Power Source	Sensors	Maximum Load	Size	Weight
Air pressure	Limit switch	Dependent on specification	1280 × 150 mm φ	

NOTE

1. For conveying both the blower-type suction jaw that collects pellets made by a powder molding press and the pellet dish.
2. Working range of the arm is 150 mm up-down and 500 mm back-forth; speed is 300 and 500 mm/s, respectively.

FREE TYPE: SUCTION   135

GRIPPER FOR BOWLING BALLS — Showa Kuatsuki Co., Ltd.

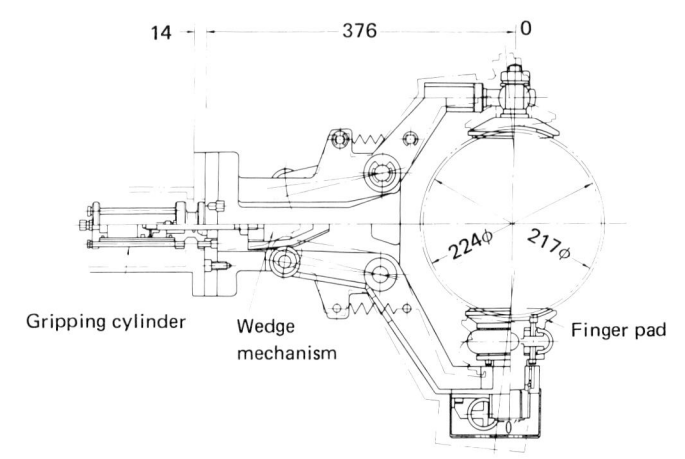

Gripping cylinder — Wedge mechanism — Finger pad

Working Speed			Working Range		
finger	wrist		finger	wrist	
open & shut	up & down	rotation	open & shut	up & down	rotation
10°/s		90°/s	10°		90°

Power Source	Sensors	Maximum Load	Size	Weight
Air pressure	None	10 kg	Sphere 217 mm φ (work)	700 kg (total robot)

NOTE

1. Gripper of the robot used to set bowling ball in a cutting machine. Wedge mechanism used for constant grasping force and amplification ability. Wrist can rotate 90°.
2. Order: sequence control by a pin board matrix.
3. Position: mechanical stopper.
4. Time: timer.
5. Velocity: preset method.

VACUUM HAND	Okamura Mfg. Co., Ltd.

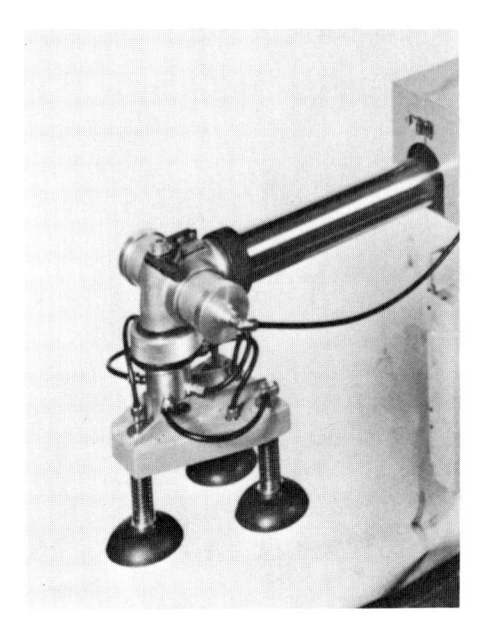

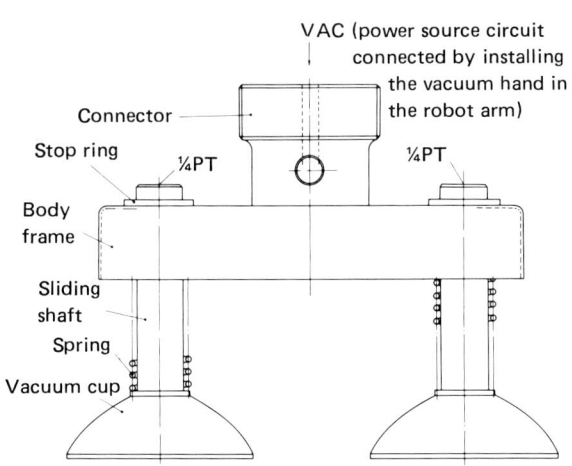

Labels on diagram: VAC (power source circuit connected by installing the vacuum hand in the robot arm), Connector, Stop ring, ¼PT, ¼PT, Body frame, Sliding shaft, Spring, Vacuum cup

Working Speed			Working Range		
finger	wrist		finger	wrist	
open & shut	up & down	rotation	open & shut	up & down	rotation
0.3 s	180°/s			0–180°	

Power Source	Sensors	Maximum Load	Size	Weight
Air pressure	Vacuum switch	5 kg	Dependent on work	

NOTE
1. Arm stroke 700 mm back-forth.
2. Three points can be freely located by moving stoppers within 700 mm.
3. Vacuum source for the hand is set in the robot.
4. Fixed sequence control by a program pin board.
5. Working range is 130 mm up-down and 220° right-left; speed is 150 mm/s and 120°/s, respectively.

FREE TYPE: MAGNETISM

MAGNETIC HAND

Daikin Kogyo Co., Ltd.

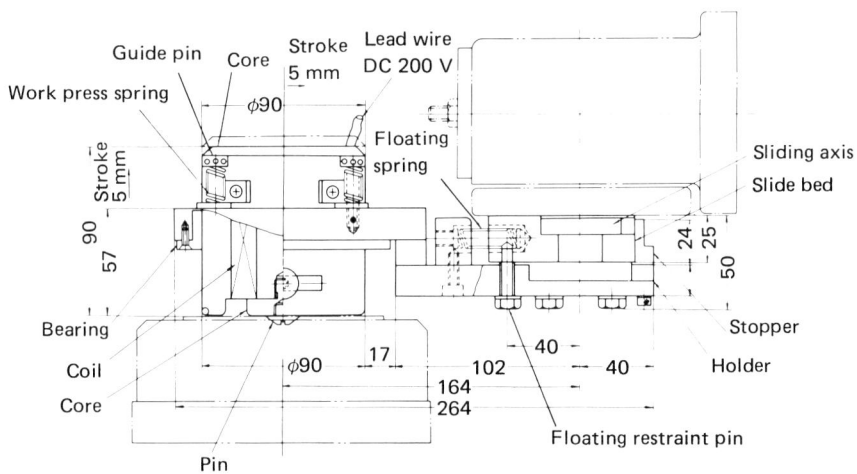

Working Speed			Working Range		
finger	wrist		finger	wrist	
open & shut	up & down	rotation	open & shut	up & down	rotation
0.3 s	90°/s	90°/s		Freely designable	

Power Source	Sensors	Maximum Load	Size	Weight
Electricity (AC 200 V, rectified)	None	10 kg (total robot)	90 mm ϕ (magnet)	8 kg

NOTE

1. Hand is equipped with both mechanism for pressing work to the jig and floating mechanism that allows more or less discrepancy between the work and the jig.
2. Magnet designed in case coil burns out and residual magnetism remains; floating spring is designed in consideration of the inertia of the work and the magnet.
3. Working range of the arm is 500 mm up-down, 400 mm back-forth, and 60° right-left; speed is 500 mm/s, 500 mm/s, and 60°/s, respectively.
4. Memory regenerative servo control in the robot.
5. On-off control in the finger.
6. Absorption force 38 kg.

MAGNETIC HAND

Daikin Kogyo Co., Ltd.

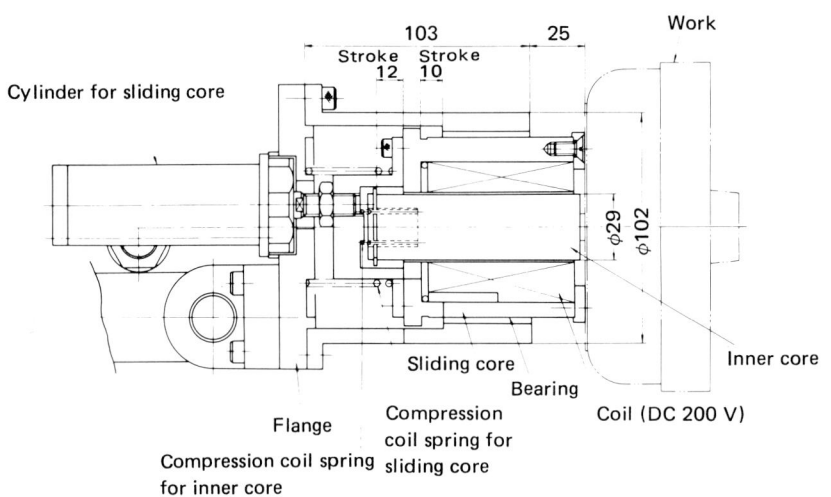

Dimensions and labels on diagram:
- 103, 25, Work
- Stroke 12, Stroke 10
- Cylinder for sliding core
- φ29, φ102
- Sliding core
- Bearing
- Inner core
- Flange
- Compression coil spring for sliding core
- Coil (DC 200 V)
- Compression coil spring for inner core

Working Speed			Working Range		
finger	wrist		finger	wrist	
open & shut	up & down	rotation	open & shut	up & down	rotation
	Freely settable		Dependent on work	Freely designable	

Power Source	Sensors	Maximum Load	Size	Weight
Oil pressure	None	5 kg		8 kg

NOTE

1. For bringing work into close contact with both vertical and horizontal sides of the jig.
2. Spring that presses work to the jig and spring that lets the work and jig fall by gravity are carefully designed.

FREE TYPE: MAGNETISM

HAND FOR TRANSFERRING ROTORS

Sumitomo Heavy Industries, Ltd.

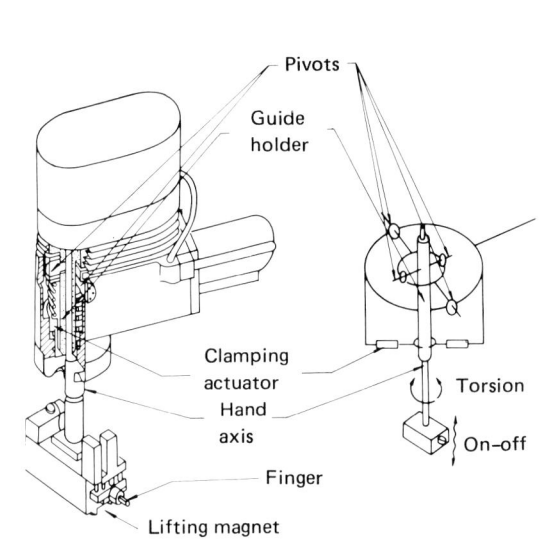

Working Speed			Working Range		
finger	wrist		finger	wrist	
open & shut	up & down	rotation	open & shut	up & down	rotation
450 mm/s		168°/s	45 mm	Freely designable	

Power Source	Sensors	Maximum Load	Size	Weight
Oil pressure (70 kg/cm^2)	Microswitch	30 kg	200 mm ϕ (lifting magnet)	10 kg

NOTE

1. Program control.
2. On-off control in the finger.
3. Absorption of rotors by a lifting magnet.
4. Working range of the arm is 500-1200 mm up-down, ±90° right-left, and 465-1305 mm back-forth; speed is 800 mm/s, 90°/s, and 800 mm/s, respectively.

MODULAR-TYPE MAGNETIC HAND (TYPE KMR-2)

Kayaba Industry Co., Ltd.

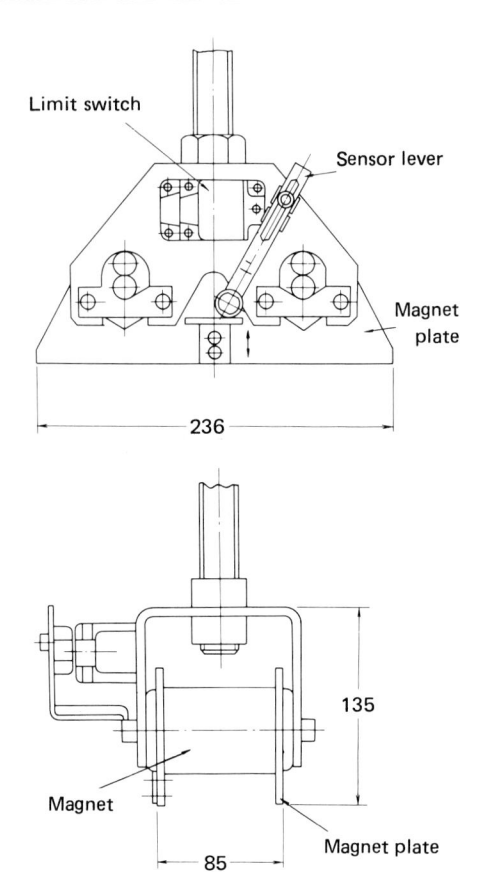

Working Speed			Working Range		
finger	wrist		finger	wrist	
open & shut	up & down	rotation	open & shut	up & down	rotation
	Freely settable			Freely designable	

Power Source	Sensors	Maximum Load	Size	Weight
Oil pressure (100 kg/cm^2), electricity (AC 200 V, 3.7 kW)	None	40 kg	1000 (L) × 600 (W) × 1485 (H) mm	480 kg

NOTE

1. Working range of the arm is 250 mm up-down, 500 mm back-forth, and 240° right-left; speed is 1.0 s, 1.5 s, and 90°/1.5 s, respectively.
   - Control: point to point control.
   - Position sensor: potentiometers.
   - Order memory: pin board matrix.
   - Position memory: potentiometers.
   - Steps: 32.
   - Location: electricity-oil pressure servomotor.
   - External synchronizing signal: receiving 4 × transmitting 2.
   - Locating error: ±1 mm.

FREE TYPE: MAGNETISM

| MAGNETIC JAW | Keiaisha Mfg. Co., Ltd. |

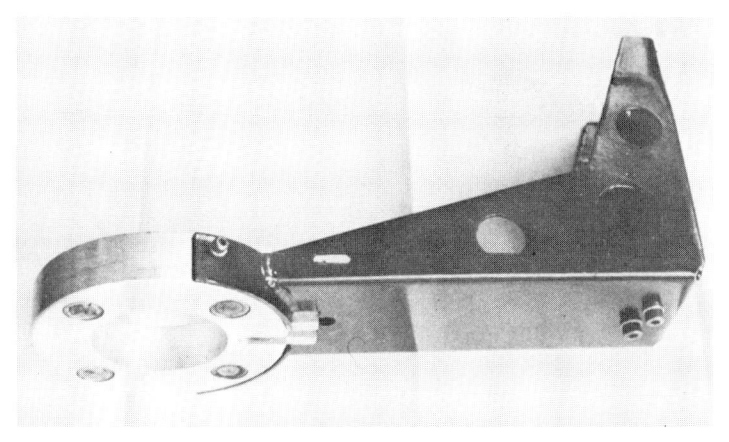

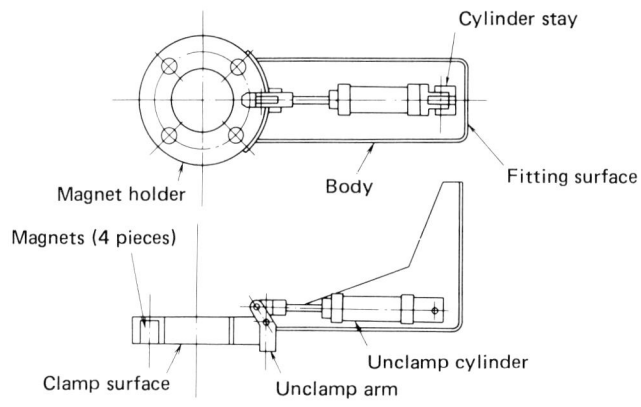

Figure labels: Cylinder stay, Fitting surface, Magnet holder, Body, Magnets (4 pieces), Clamp surface, Unclamp arm, Unclamp cylinder

Working Speed		Working Range	
finger	wrist	finger	wrist
open & shut	up & down rotation	open & shut	up & down rotation
	Freely settable		Freely designable

Power Source	Sensors	Maximum Load	Size	Weight
Air pressure	None	Dependent on work	Freely designable	4.8 kg

NOTE

1. Clamping by magnetic absorption and unclamping by the arm connected with a cylinder.
2. Mainly used for transferring sheet materials. (The figure above shows an unclamping phase.)

| INNER GRASPING SWING LEVER HAND | Yasukawa Electric Mfg. Co., Ltd. |

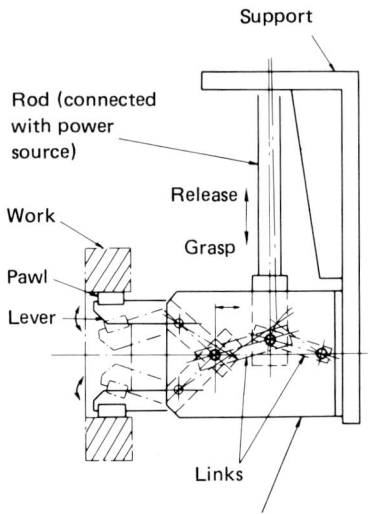

Working Speed				Working Range			
finger	wrist			finger	wrist		
open & shut	up & down		rotation	open & shut	up & down		rotation
1 s (1.5 s standard timer setting)	Freely settable			About 50 mm	Freely settable		
Power Source	Sensors		Maximum Load		Size		Weight
Electricity (motors)	None		About 10 kg		Inner diameter about 50 mm ϕ (work)		About 6 kg

NOTE

1. When the other machine hands over a work from the chuck that grasps it from the outside, the hand holds it from the inside.
2. To satisfy the size constraint in the work axis direction, the hand is shortened by arranging the power source vertically to the pawl opening direction.
3. On-off control by timer and limit switch.

FREE TYPE: OTHERS

WIPING HAND	Toshiba Co.

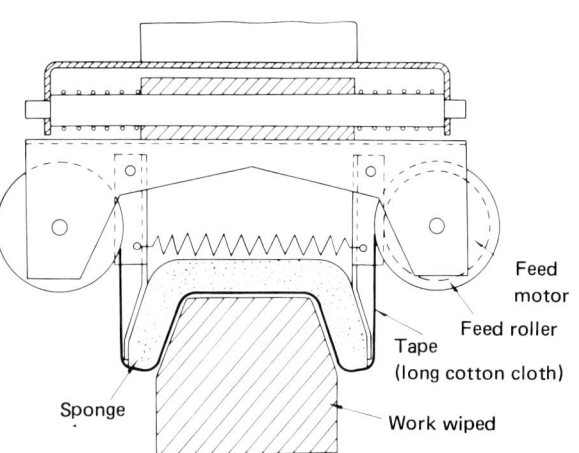

Feed motor
Feed roller
Tape (long cotton cloth)
Sponge
Work wiped

Working Speed		Working Range		
finger	wrist		finger	wrist	
open & shut	up & down	rotation	open & shut	up & down	rotation
	90°/s	90°/s		220°	180°

Power Source	Sensors	Maximum Load	Size	Weight
Air pressure	None	10 kg		4 kg

NOTE

1. For automation of wiping objects with various cross sections. Even point-to-point-type robots can use it. The tape-like wiping cloth need not be frequently exchanged.
2. Can wipe objects with various cross sections.
3. Can wipe even if the wiping locus has curvature.
4. Use of tape-like wiping cloth reduces the frequency of cloth exchange.
5. 10 m wiping length.

HAND FOR WORKING ON REAR HEAD OIL HOLES	Daikin Kogyo Co., Ltd.

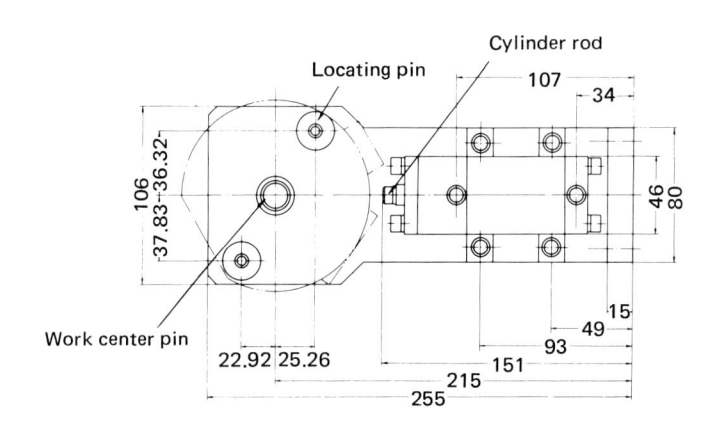

Cylinder rod
Locating pin
Work center pin

Working Speed		Working Range	
finger	wrist	finger	wrist
open & shut	up & down rotation	open & shut	up & down rotation
0.4 s	180°/s		Freely designable

Power Source	Sensors	Maximum Load	Size	Weight
Air pressure (4 kg/cm² G)	None	15 kg (total robot)	255 × 106 mm	5.7 kg

NOTE

1. Objects precisely located by the pins in the hand and set in a small jig, not grasped from the outside.
2. The cylinder diameter is designed according to the friction coefficient and the acceleration of objects.
3. Reliable grasping consists of setting an object by a work hole center pin, fixing it by a locating pin, and pressing it by a cylinder rod.
4. Working range of the arm is 100 mm up-down, 520 mm back-forth, and 60° right-left; speed is 500 mm/s (both) and 60°/s, respectively.
5. Fixed sequence control in the robot and on–off control in the finger.
6. Grasping force 24 kg.

FREE TYPE: OTHERS

GRIPPER FOR PRESS LOADING

Showa Kuatsuki Co., Ltd.

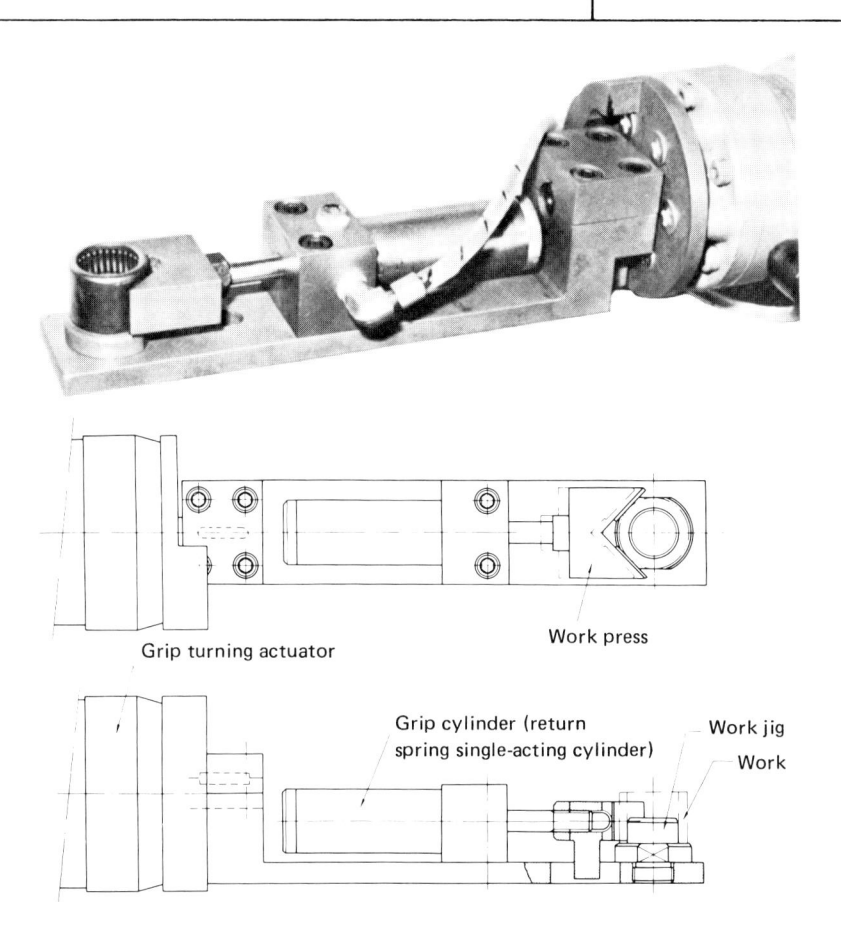

Grip turning actuator — Work press

Grip cylinder (return spring single-acting cylinder) — Work jig — Work

Working Speed			Working Range		
finger	wrist		finger	wrist	
open & shut	up & down	rotation	open & shut	up & down	rotation
20 mm/s	180°/s		10 mm	180°	

Power Source	Sensors	Maximum Load	Size	Weight
Air pressure	None	26 kg	26 mm ϕ × 20 mm ϕ × 20 mm (thickness)	

NOTE

1. Gripper of robot for press loading with return spring single-acting cylinder. The gripped work is not released in case of accident. Objects are cylindrical and can be rotated 180° to put them on the jig, clamp them, and supply them after reversing.
2. Order: fixed sequence control.
3. Position: mechanical stopper.
4. Time: timer.
5. Velocity: preset method.

SWING HOOK	Marol Company, Ltd.

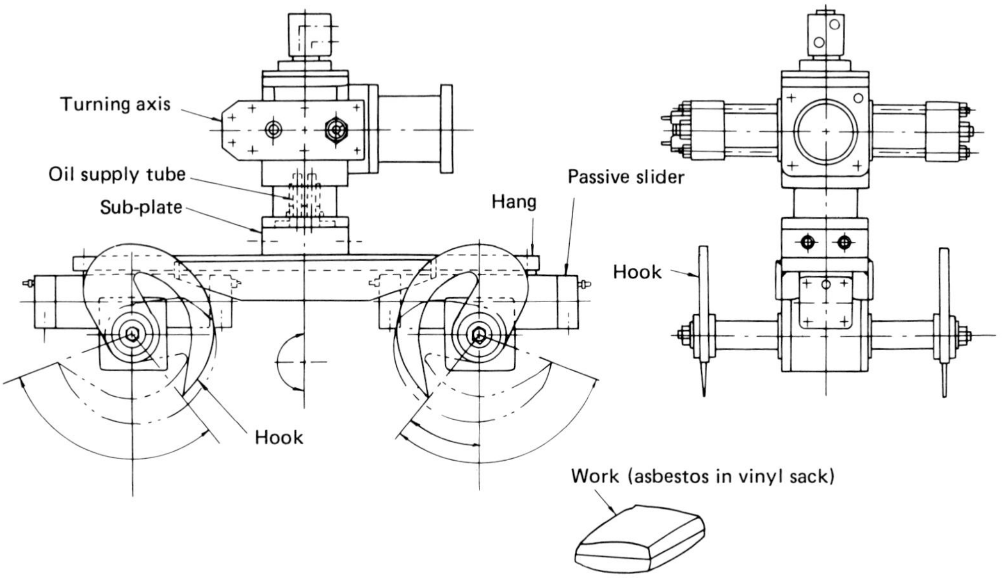
Turning axis
Oil supply tube
Sub-plate
Hang
Passive slider
Hook
Hook
Work (asbestos in vinyl sack)

Working Speed			Working Range		
finger	wrist		finger	wrist	
open & shut	up & down	rotation	open & shut	up & down	rotation
Free		4 s/90°	90°	90°	180°

Power Source	Sensors	Maximum Load	Size	Weight
Oil pressure	None	50 kg		80 kg

NOTE

1. Hand for hooking sacks.
2. 10 kg torque at the hooking axes, and about 11 kg in the connected direction.

HAND FOR AUTOMATIC ARC WELDING

Kawasaki Heavy Industries, Ltd.

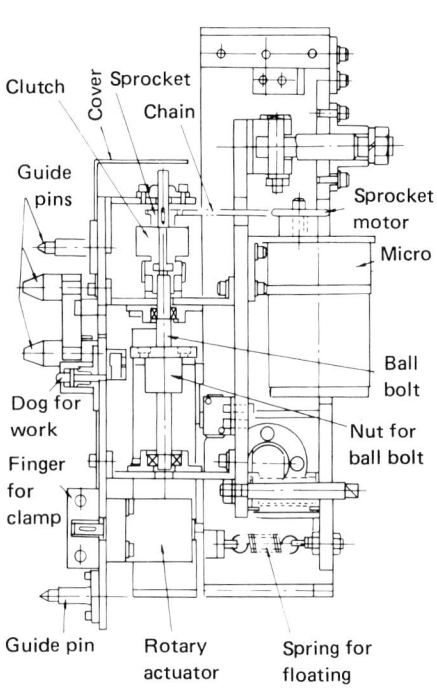

Labels on diagram: Clutch, Cover, Sprocket, Chain, Guide pins, Sprocket motor, Micro, Ball bolt, Nut for ball bolt, Dog for work, Finger for clamp, Guide pin, Rotary actuator, Spring for floating

Working Speed			Working Range		
finger	wrist		finger	wrist	
open & shut	up & down	rotation	open & shut	up & down	rotation
	Freely settable			Freely designable	

Power Source	Sensors	Maximum Load	Size	Weight
Air pressure and electricity	None	1–2 kg		18 kg

NOTE

1. Kawasaki Unimate (a robot) with this hand takes an object, holds it in the center chucks, presses it against another object, and then does arc welding by letting the arc welding torch run automatically.
2. Point to point control robot.
3. The hand has a clamping mechanism, mechanism for imitating the other work, and mechanism for transferring the welding torch, so it needs no work press jig.
4. The hand has checking mechanisms (by air pressure), torch drive mechanism (by micromotors), and floating mechanism (by springs).

WELDING GUN AUTOCHANGER

Dendensha Mfg. Co., Ltd.

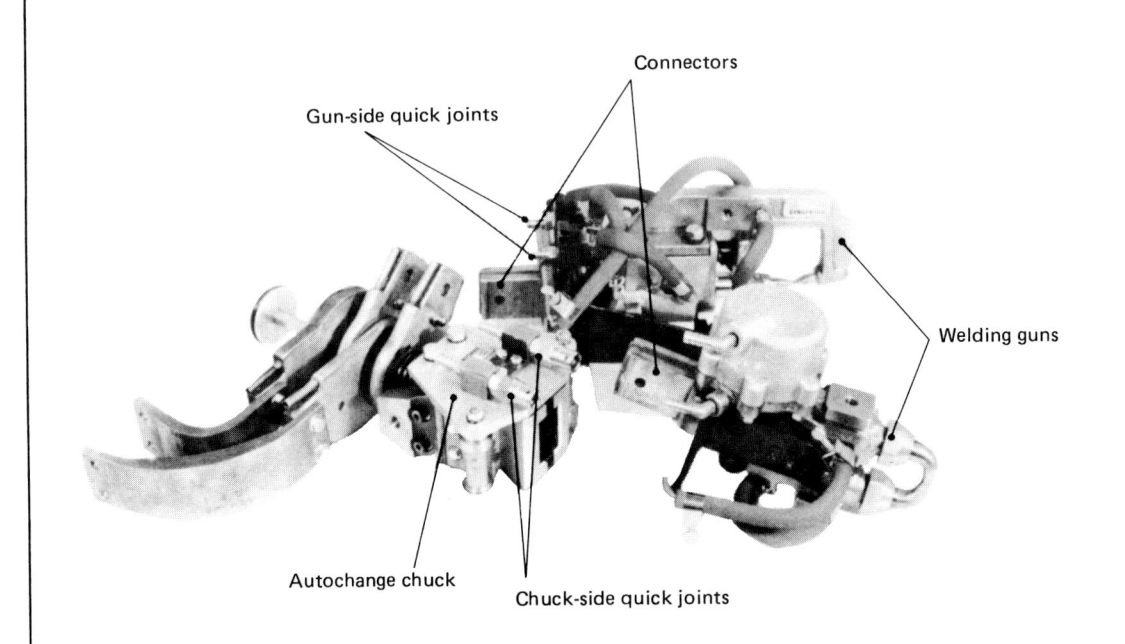

Connectors
Gun-side quick joints
Welding guns
Autochange chuck
Chuck-side quick joints

Working Speed			Working Range		
finger	wrist		finger	wrist	
open & shut	up & down	rotation	open & shut	up & down	rotation
	Freely settable			Freely designable	

Power Source	Sensors	Maximum Load	Size	Weight
Air pressure	None			

NOTE

1. When spot welding, the robot has the welding gun of a portable spot welding machine. The welding gun may need to be changed for other welding, and usually two gun sets are prepared in manual work.
2. The robot replaces a human worker, so the gun must be changed automatically. This hand has been developed for that purpose, and the connecting terminal is used to connect electricity, air, and water needed for welding.

FREE TYPE: OTHERS

HAND FOR ARC WELDING

Dendensha Mfg. Co., Ltd.

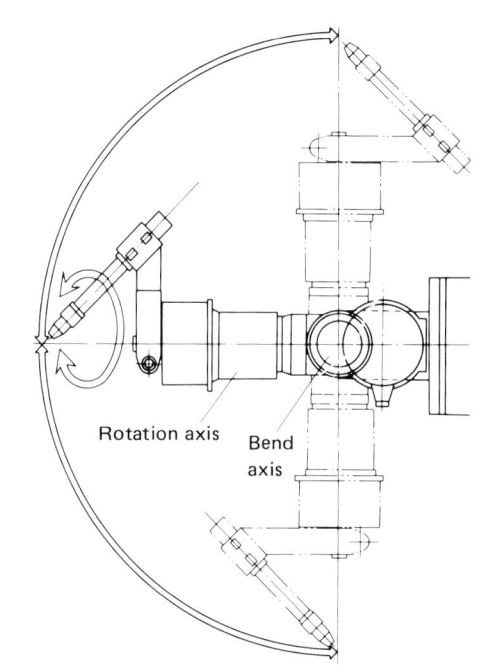

Rotation axis Bend axis

Working Speed			Working Range		
finger	wrist		finger	wrist	
open & shut	up & down	rotation	open & shut	up & down	rotation
	20°/s	30°/s		±90°	±150°

Power Source	Sensors	Maximum Load	Size	Weight
Electricity (pulse motor)	None		120 × 120 × 160 mm	

NOTE

1. Hand developed as a wrist for the arc welding robot, so the pulse motor used as an actuator is included and compact.
2. The hand with only the rotation axis can be used as the wrist for arc welding (see photograph above); addition of the bend makes 2 degrees of freedom possible.

Part 3

PATENTS

The following pages show typical examples of patents and utility models arranged chronologically according to the same classification used in Part II. The functions of these devices are indicated by the titles and sketches.

PLURAL PICK-UP HAND

Application No.	PAT. 1973-78655
Applicant	Hitachi

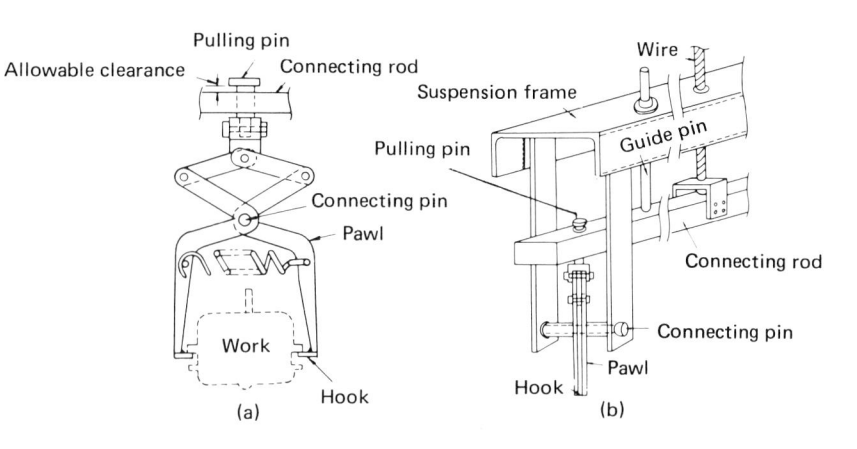

HAND WITH INDEPENDENT PAWLS

Application No.	913266
Applicant	Aida Engineering

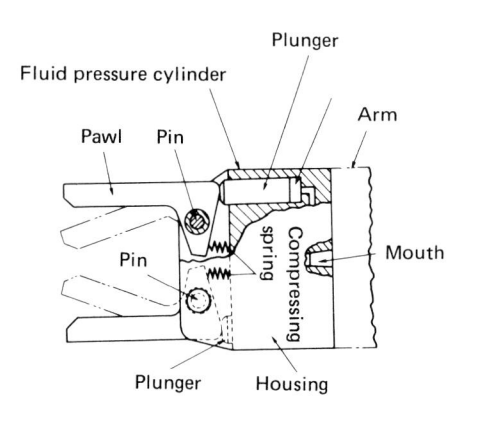

THREE-PAWL HAND

Application No.	PAT. 1974-32352
Applicant	Fanuc

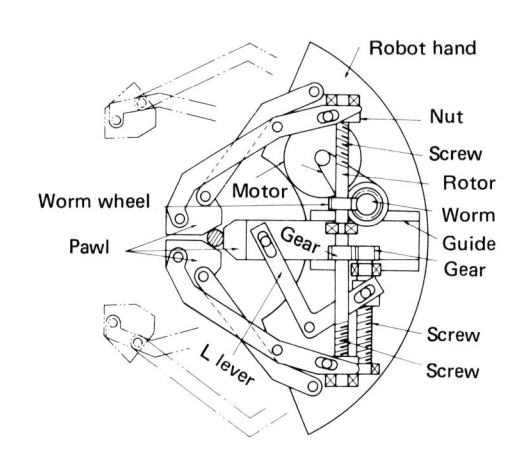

FINGERS WITH WORK SENSOR

Application No.	PAT. 1974-72864
Applicant	Hitachi

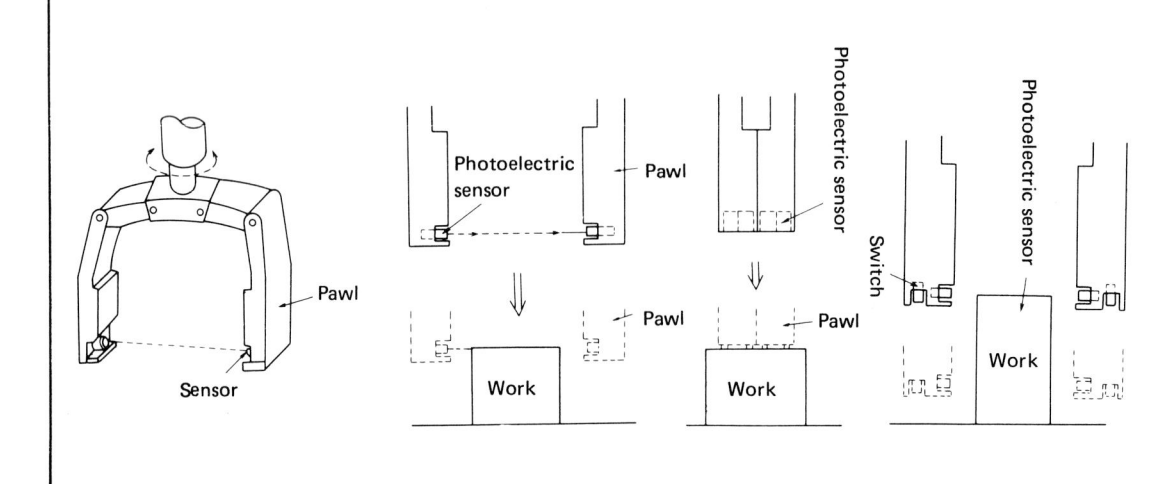

GRASP HAND FOR MULTIPLE WORKS ON CONVEYOR

Application No.	PAT. 1975-8254
Applicant	Hitachi

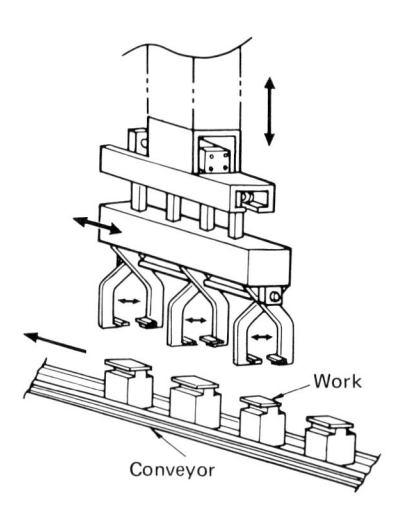

GRASP FINGERS FOR LARGE-SCALE BODIES

Application No.	PAT. 975616
Applicant	Toyota Motor

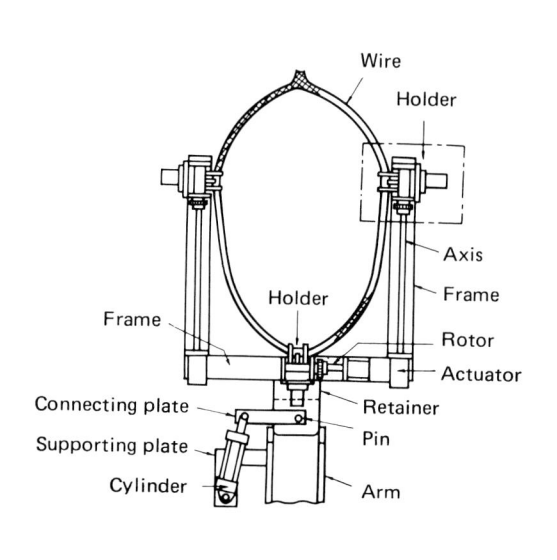

FINGERS WITH THE ABILITY TO RECOGNIZE WORK

Application No.	PAT. 1975-15259
Applicant	Tokico

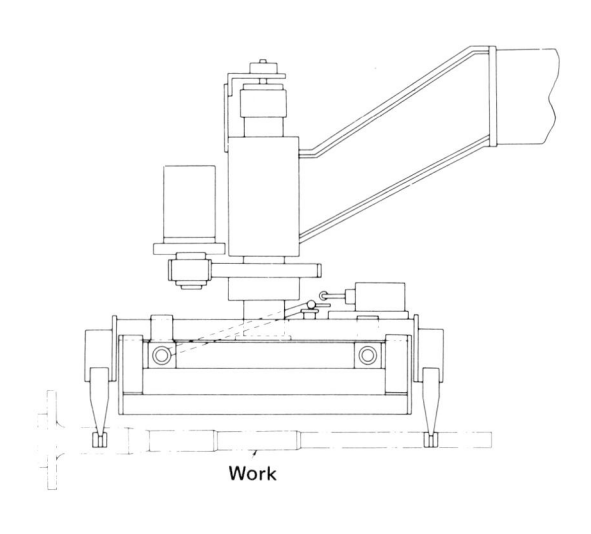

HAND WITH A COUPLE OF CLAMP ARMS

Application No.	PAT. 1975-95965
Applicant	Komatsu

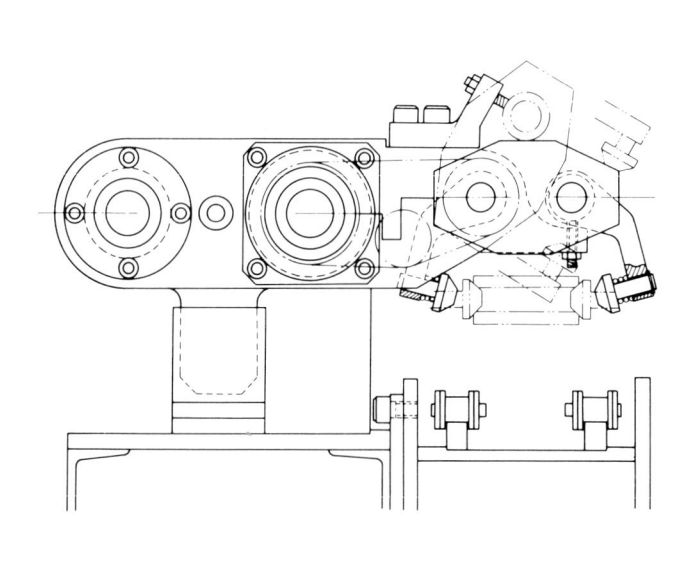

FREE TYPE: PICKING

HAND WITH THREE CONCENTRIC FINGERS

Application No.	PAT. 1976-12559
Applicant	Yamatake Honeywell

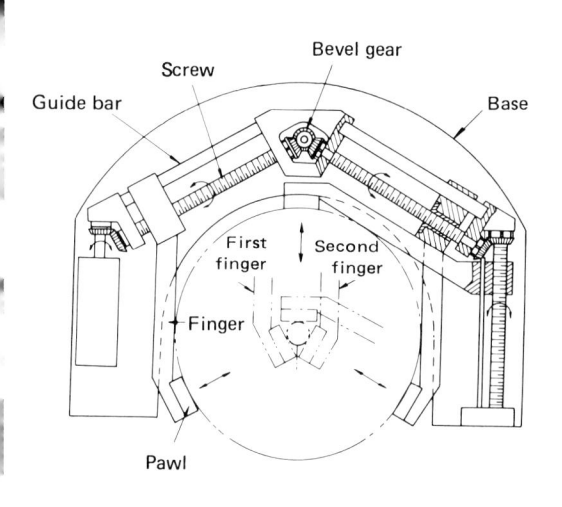

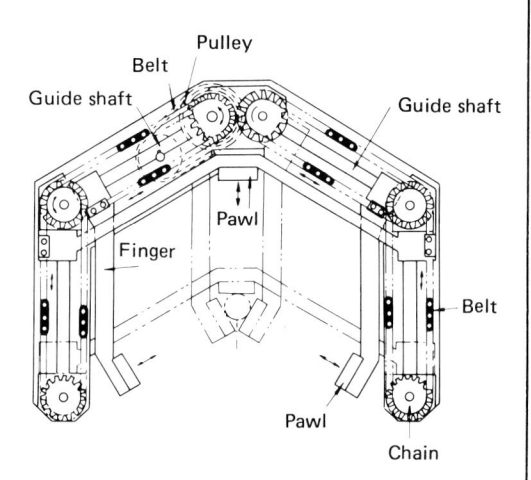

GRASPING MANIPULATOR FOR PLURAL GOODS

Application No.	PAT. 1976-23953
Applicant	Matsushita Elec.

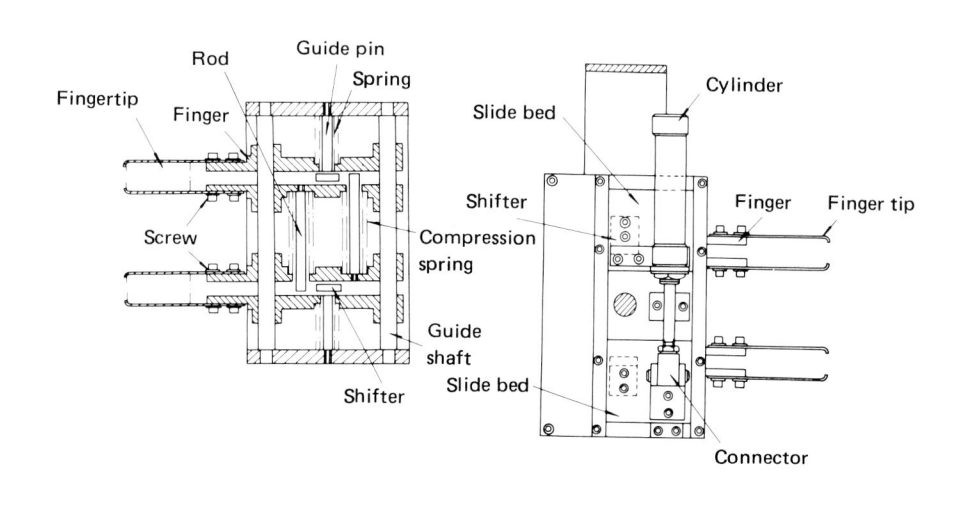

HAND POWERED IN COMMON WITH THE TRANSFER ARM

Application No.	PAT. 1976-112061
Applicant	Hitachi

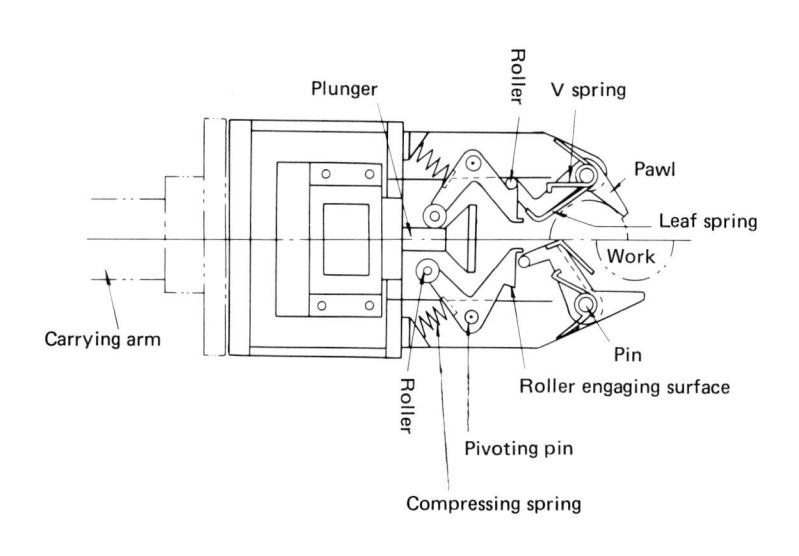

SNAPPING DEVICE

Application No.	Utility model 1974-91968
Applicant	Matsushita Elec.

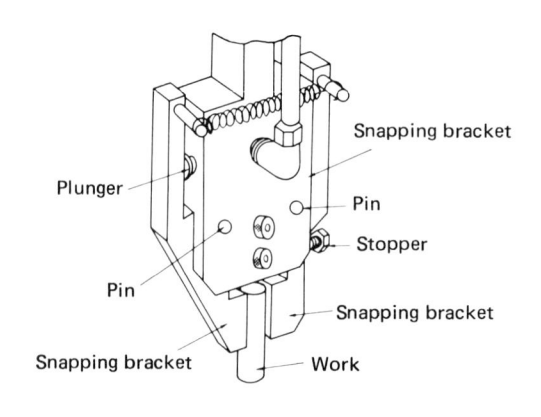

FREE TYPE: PICKING

HAND FOR SLIDING CYLINDERS IN TWO POSITIONS

Application No.	Utility model 1975-91973
Applicant	Toyoda Machine Works

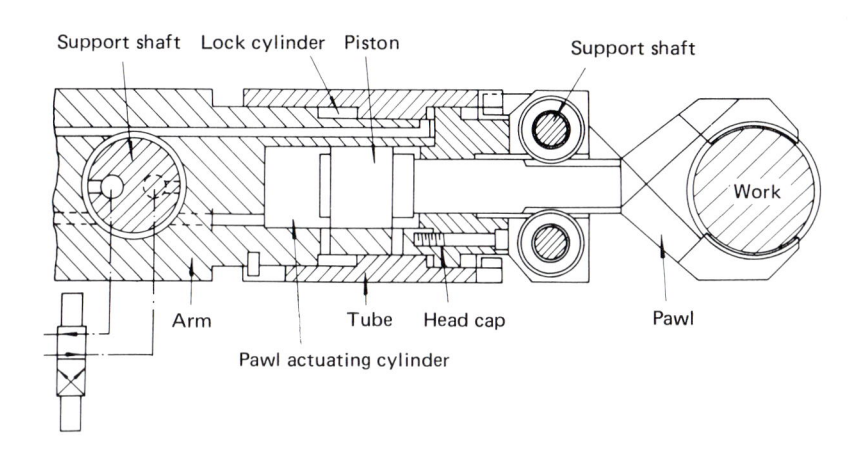

Labels: Support shaft, Lock cylinder, Piston, Support shaft, Work, Arm, Tube, Head cap, Pawl, Pawl actuating cylinder

GRASPING DEVICE FOR MATERIALS

Application No.	Utility model 1974-101469
Applicant	Toyoda Machine Works

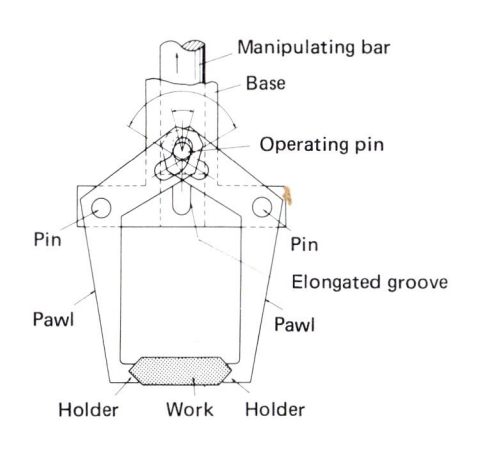

Labels: Manipulating bar, Base, Operating pin, Pin, Pin, Elongated groove, Pawl, Pawl, Holder, Work, Holder

HAND WITH A LEAF SPRING SWITCH MECHANISM

Application No.	Utility model 1975-115941
Applicant	Li Euan He

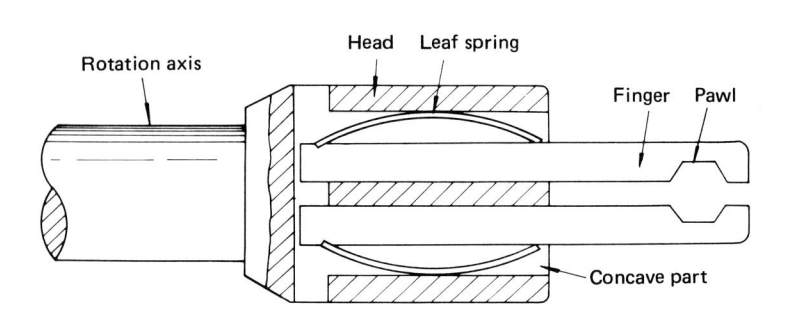

TRANSFER PAWLS FOR NUTS

Application No.	Utility model 1975-115943
Applicant	Hoichi Hammura

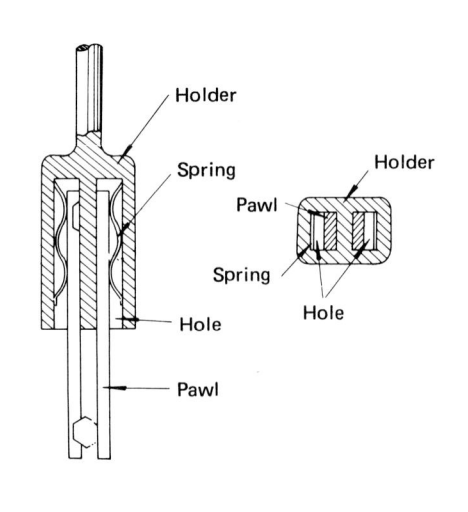

TRANSFER DEVICE IN THE RIGHT AND LEFT, UP AND DOWN DIRECTIONS

Application No.	Utility model 1975-123065
Applicant	Daifuke Kiko

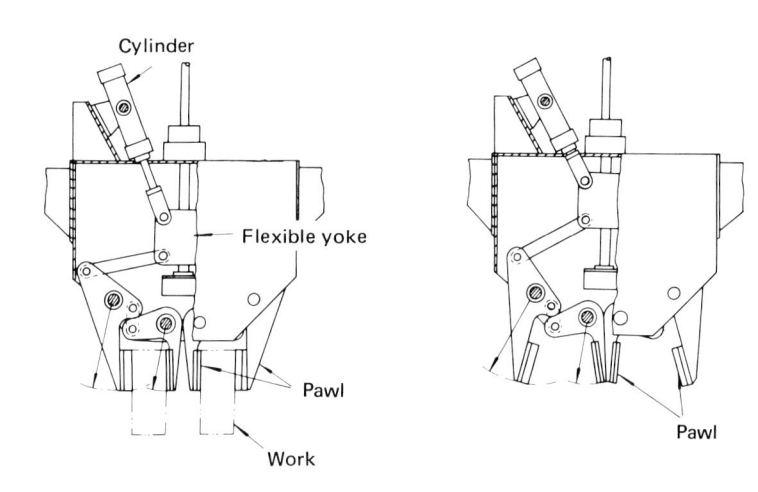

SNAP DEVICE FOR OBJECTS SUCH AS SPOONS AND DISHES

Application No.	Utility model 1975-123070
Applicant	Shibayama Kikai

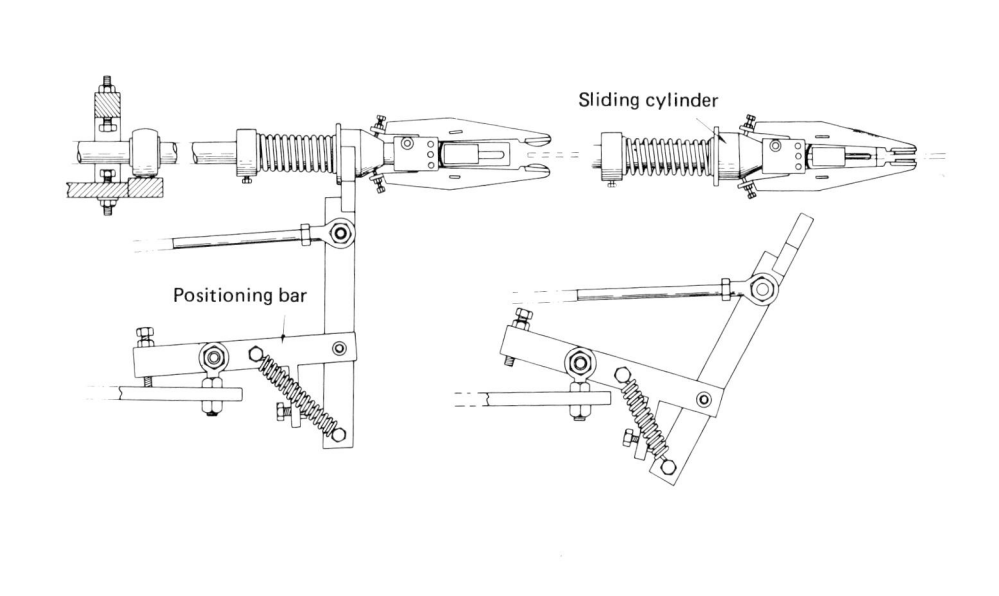

SNAP INSTRUMENT FOR CUBE SUGAR

Application No.	Utility model 1975-125957
Applicant	Marutoku Seito

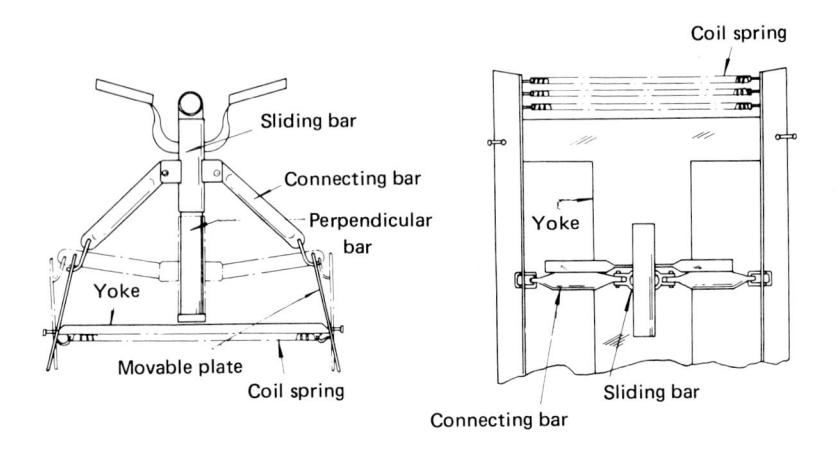

WORK GRIPPER

Application No.	Utility model 1975-127175
Applicant	Seiko Seiki

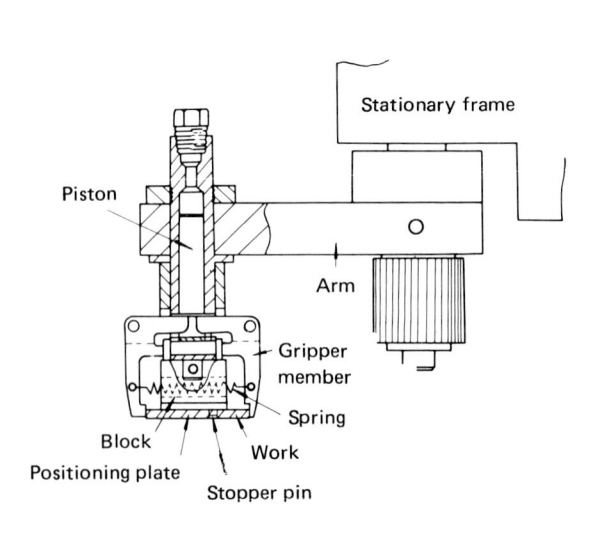

HAND WITH SNAP IN THE GUIDE FRAME

Application No.	Utility model 1975-136174
Applicant	Kazukito Machida

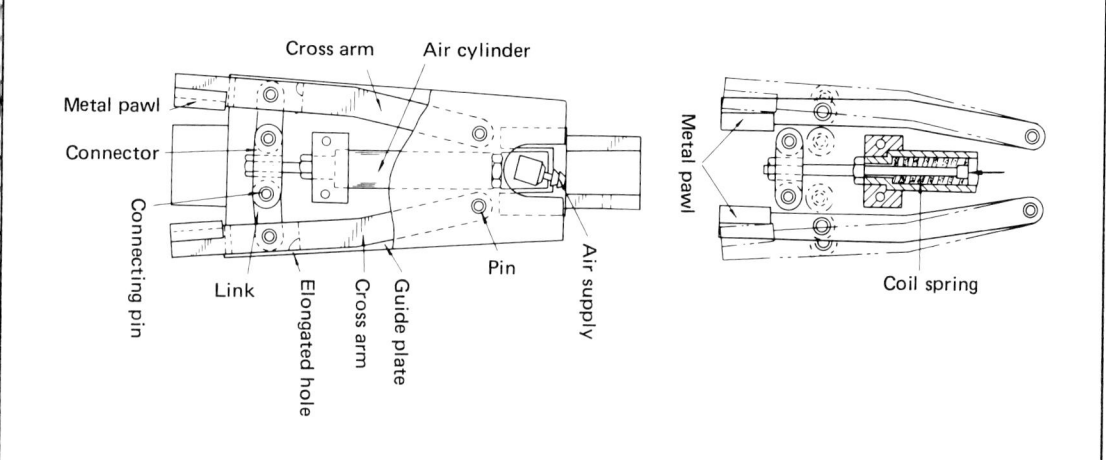

HAND WITH REMOVABLE GRASP FRAMES

Application No.	Utility model 1976-14473
Applicant	Komatsu

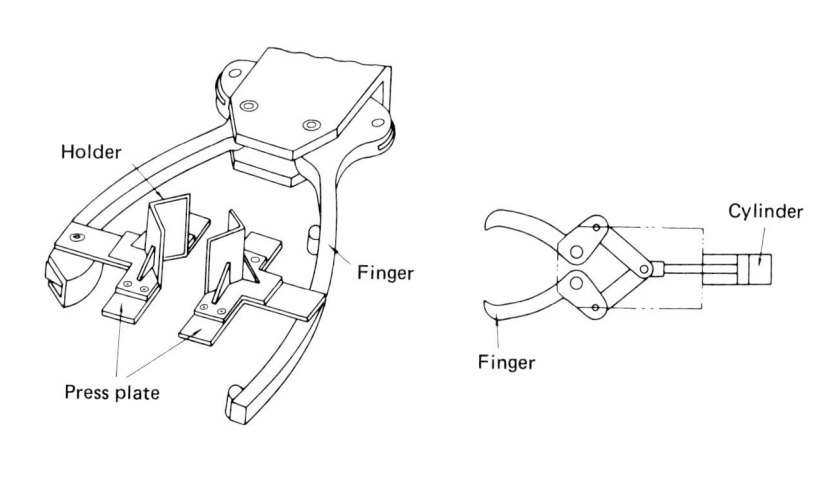

GRASPING HAND MOVABLE UP AND DOWN AND ROTATIONALLY

Application No.	Utility model 1976-69274
Applicant	Nippon Yusoki

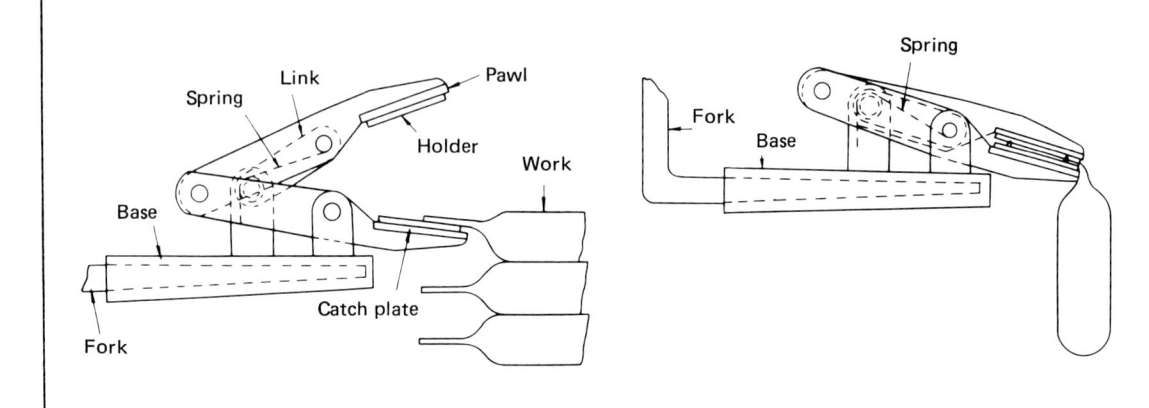

TRANSPORT HAND FOR WORK ON TURNING MACHINE TOOL

Application No.	Utility model 1976-111581
Applicant	Hitachi Seiki

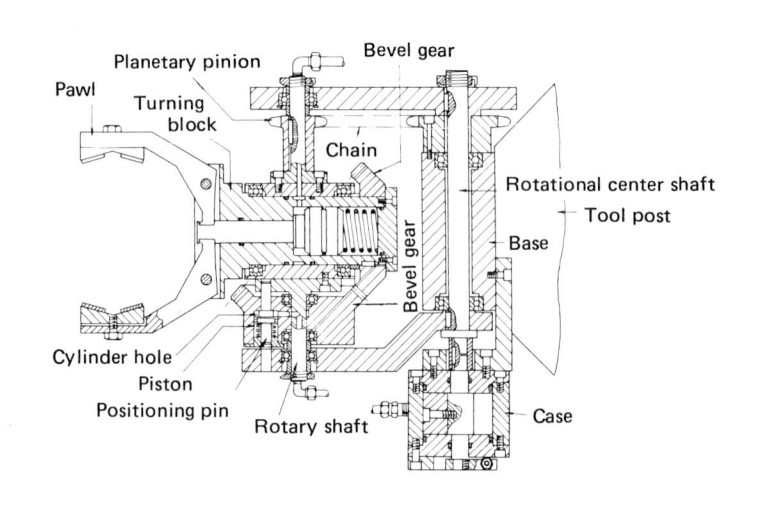

SAFE AND RELIABLE HAND

Application No.	PAT. 1972-28656
Applicant	Meidensha

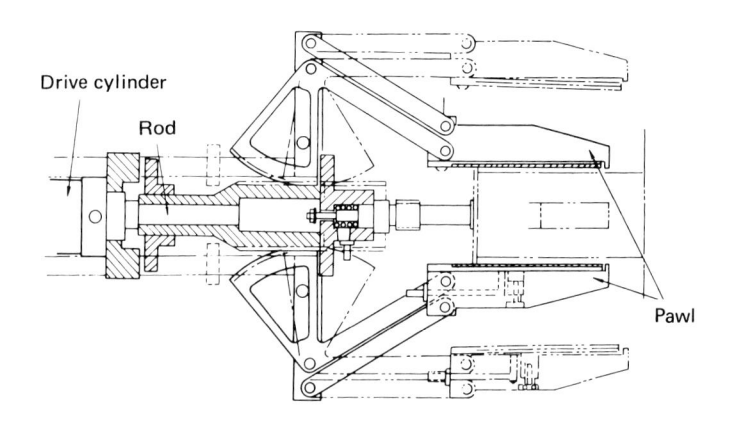

GRASPING DEVICE WITH PROBE DIRECTLY TOUCHABLE TO WORK

Application No.	PAT. 1973-49156
Applicant	Komatsu

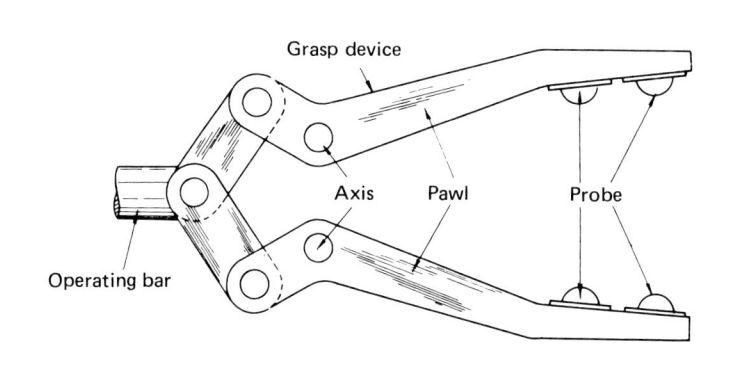

PROBE FINGER INTO GRASP PRESSURE AND DISPLACEMENT

Application No.	PAT. 1973-53464
Applicant	Tokyo Seimitsu

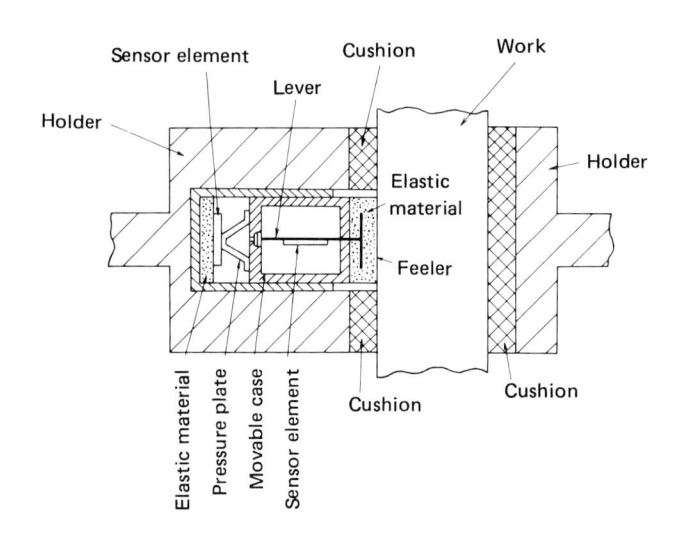

RAPIDLY CHANGEABLE CHUCK

Application No.	PAT. 1973-58561
Applicant	Hitachi

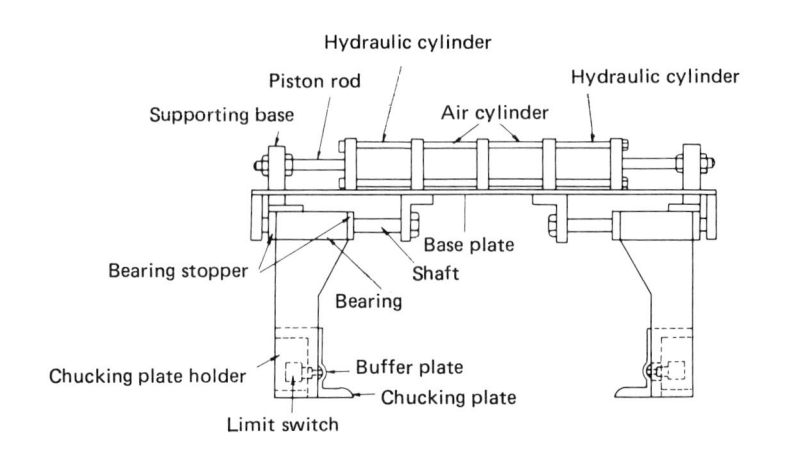

HAND FOR OBJECTS OF COMPLEX SHAPE

Application No.	PAT. 1973-65655
Applicant	Komatsu

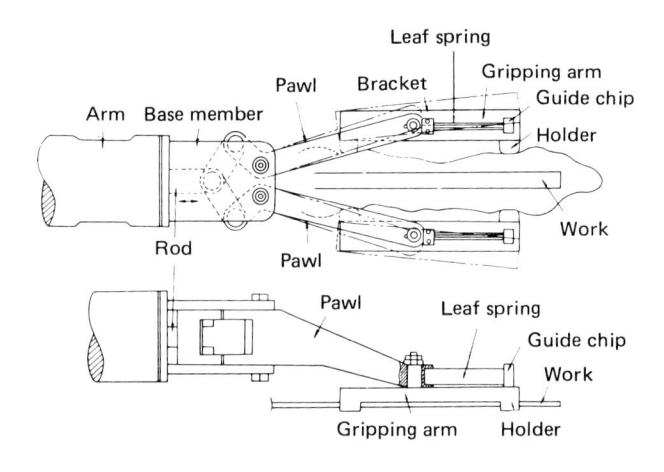

HAND WITH SNAP SENSORS

Application No.	PAT. 976131
Applicant	Tsuvakimoto Chain

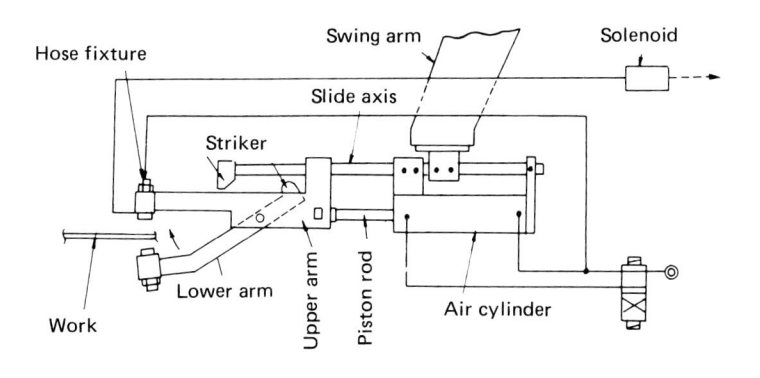

MULTIJOINT GRIPPER

Application No.	936070
Applicant	Aida Engineering

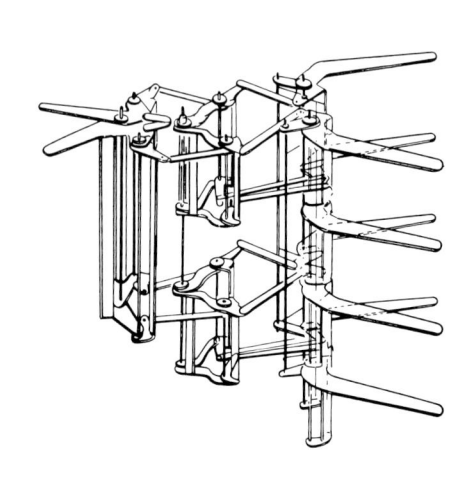

HAND ADAPTABLE TO THE WEIGHT, SHAPE, AND STIFFNESS OF MATERIALS

Application No.	PAT. 1974-3351
Applicant	Hitachi

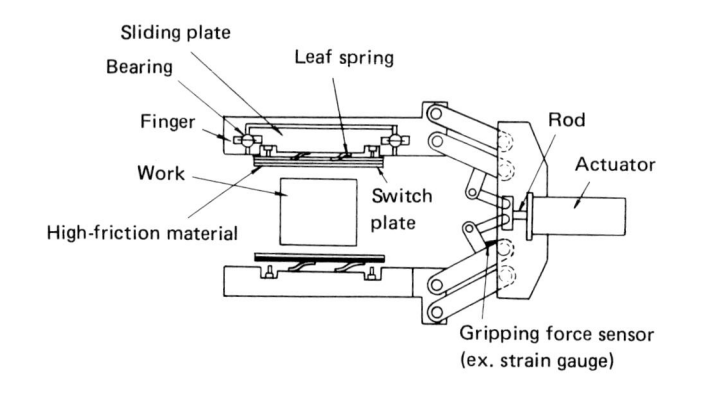

FREE TYPE: SNAP

HAND WITH CHANGEABLE GRASPING FORCE AND SPEED

| Application No. | PAT. 1974-3353 |
| Applicant | Kondo Seisakusho |

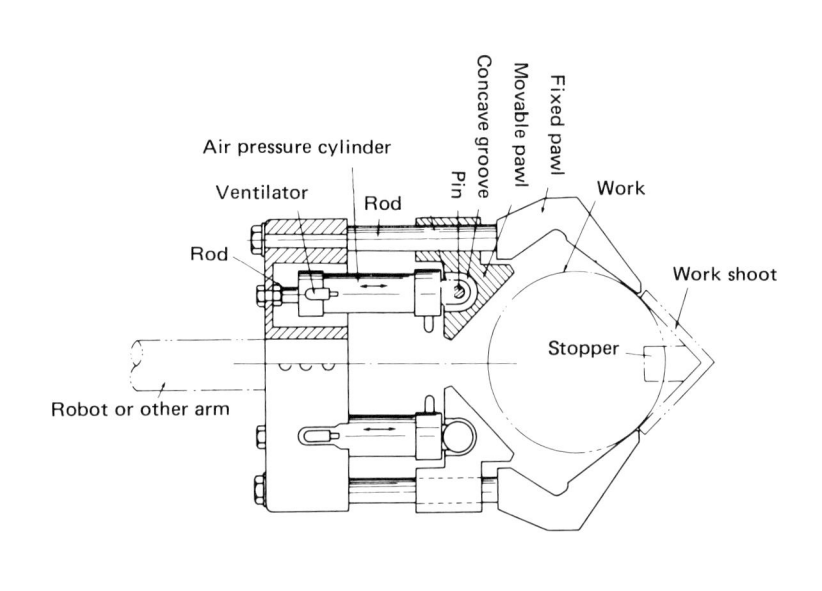

PARALLEL LINKS SWITCHED BY ONE CYLINDER

| Application No. | PAT. 1974-22705 |
| Applicant | Agency of Industrial Science and Technology |

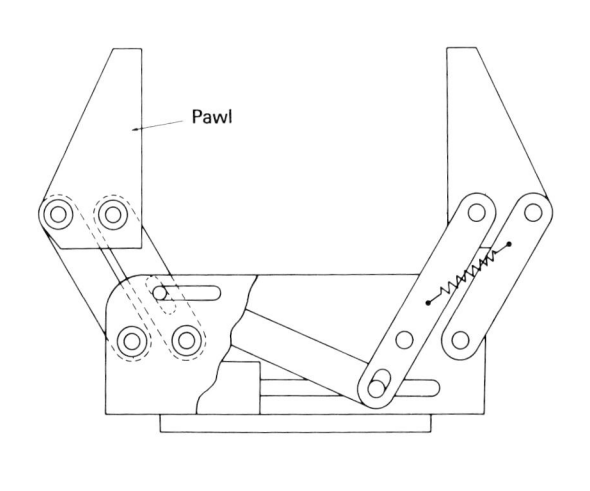

GRASPING HAND FOR NARROW SPACE

Application No.	PAT. 1974-28058
Applicant	Tokico

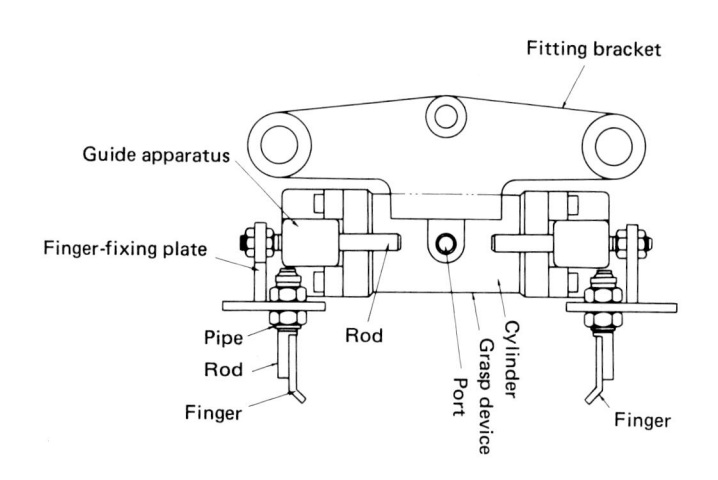

FINGERS CONFIRMING THE GRASP OF THIN PLATES

Application No.	PAT. 1974-36304
Applicant	Tadashi Aizawa

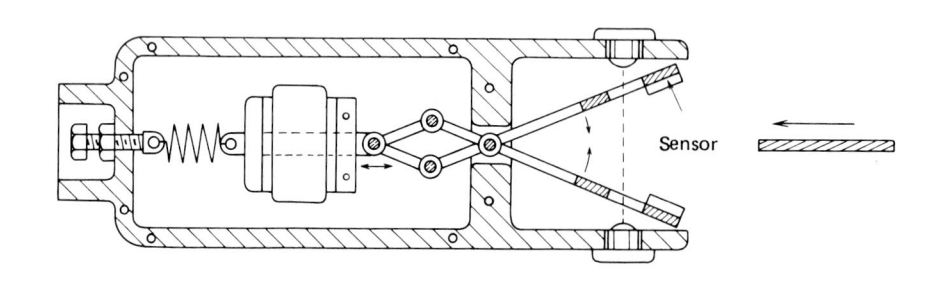

FINGERS WITH REVERSIBLE LONG AND SHORT LEVERS

Application No.	US: 3771825
Applicant	Mitsui Engineering & Shipping, Glory Kogyo

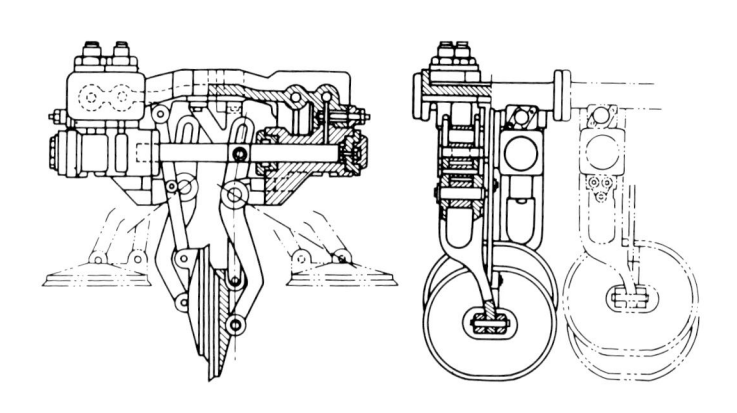

FINGER WITH CONDUCTIVE PROBES

Application No.	PAT. 1974-68456
Applicant	Hitachi

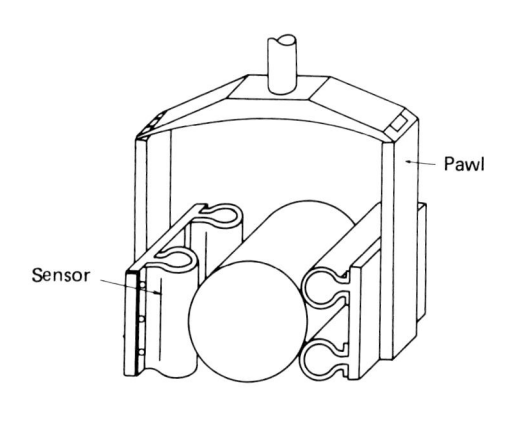

FINGERS ADJUSTABLE TO THE DISPLACEMENT OF OBJECTS

Application No.	PAT. 1974-68458
Applicant	Fujitsu

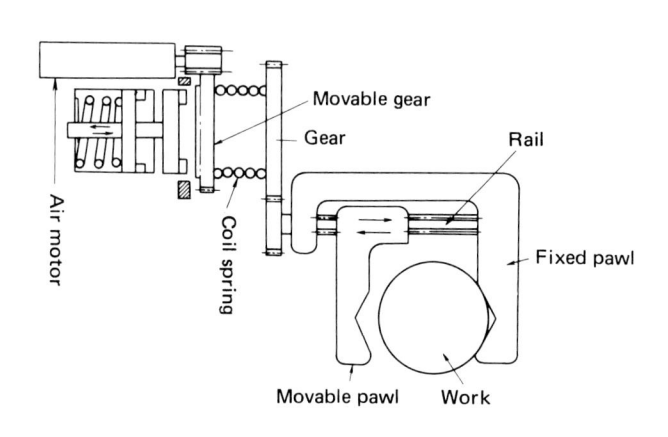

GRASPING HAND FOR CYLINDRICAL FIBER WORK

Application No.	PAT. 1974-82059
Applicant	Kodensha

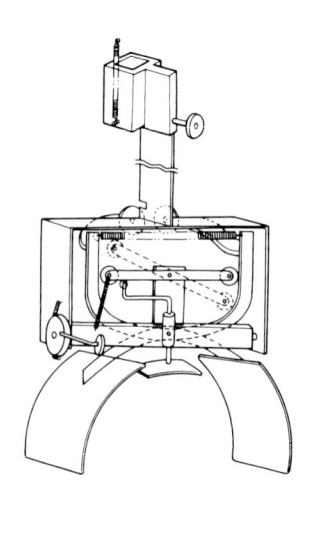

FINGER WITH RESISTANCE POWDER SENSOR

Application No.	PAT. 1974-92757
Applicant	Kawasaki Heavy Ind.

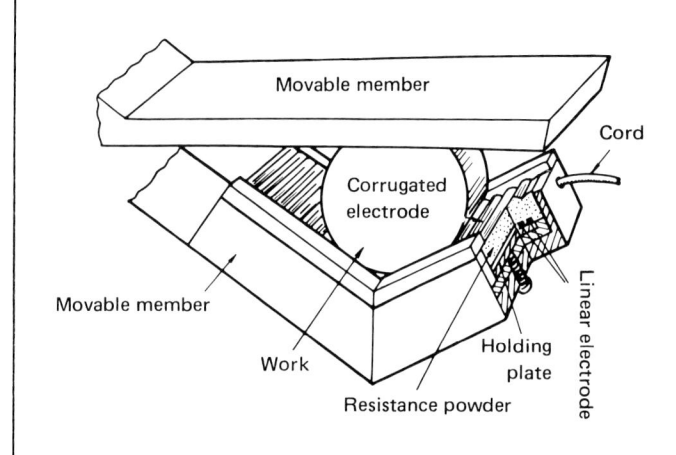

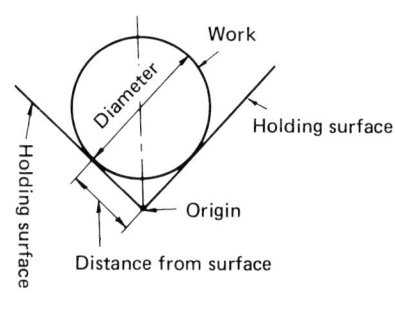

HAND WITH TWO TYPES OF FINGER

Application No.	PAT. 1974-103356
Applicant	Mitsubishi Elec.

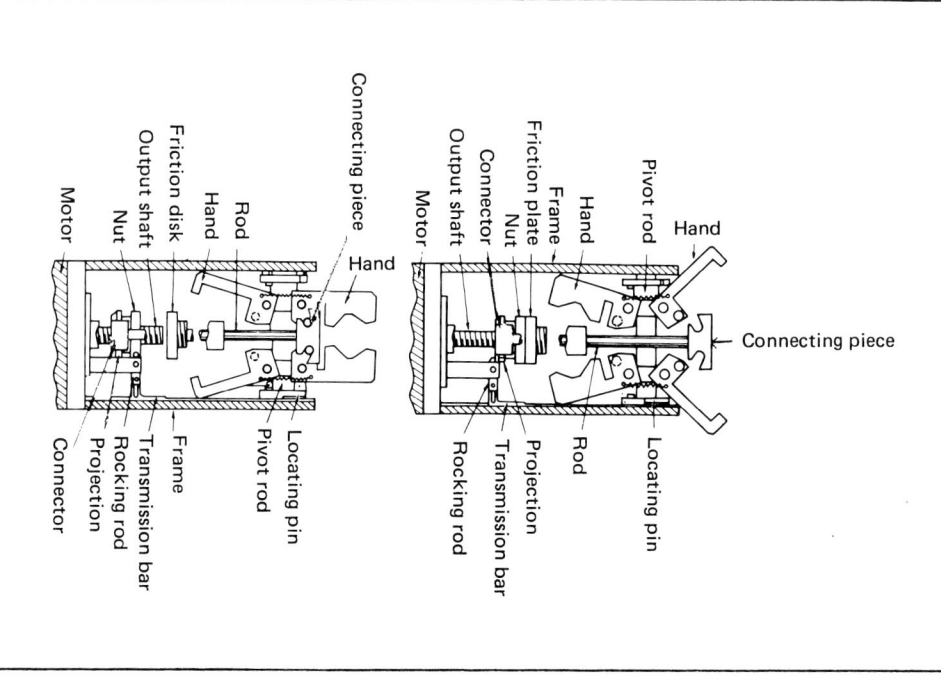

HAND FOR SETTING THE CENTER OF OBJECTS

Application No.	PAT. 1974-104355
Applicant	Hitachi

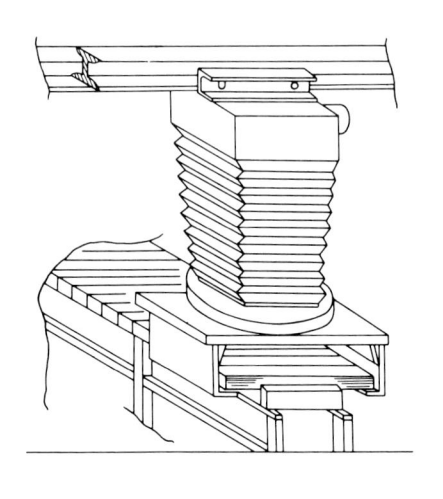

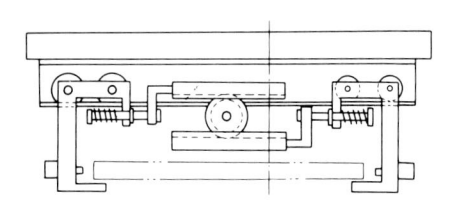

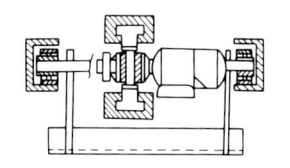

FINGER FOR GRASPING THE CENTER OF OBJECTS BY CONTACT

Application No.	PAT. 1974-104357
Applicant	Tokyo Shibaura Elec.

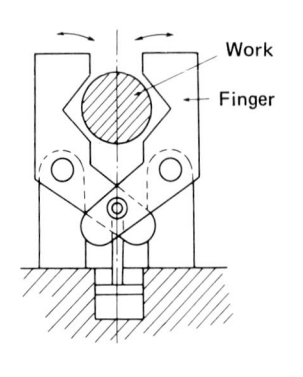

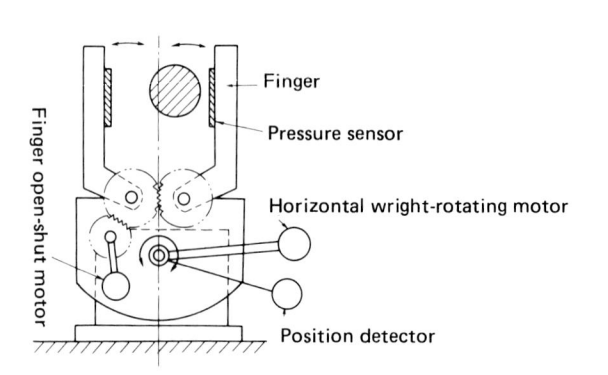

CENTER HOLDING HAND WITH SENSORS

Application No.	PAT. 1974-104358
Applicant	Tokyo Shibaura Elec.

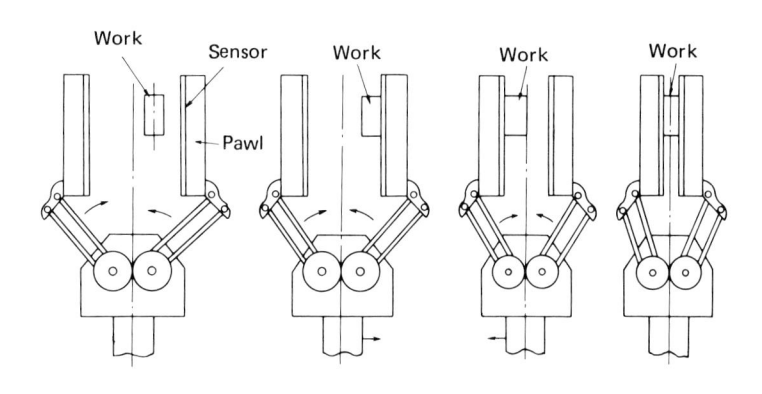

GRASPING HAND FOR NARROW SPACE

Application No.	PAT. 1974-28059
Applicant	Tokico

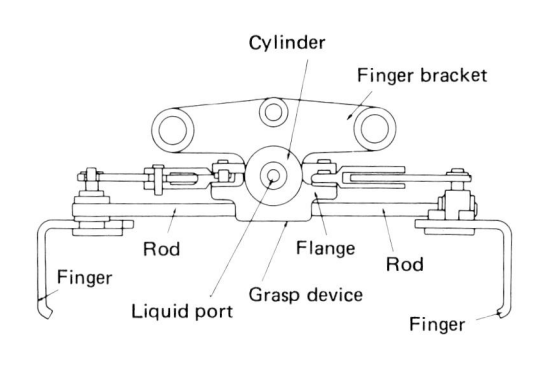

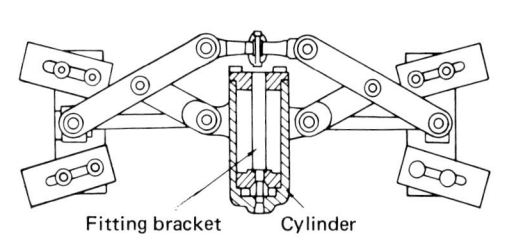

FINGER FOR MODIFYING THE DECLINATION OF OBJECTS

Application No.	US: 3963271, UK: 1481624, W. Germany: 2449078
Applicant	Yamatake Honeywell

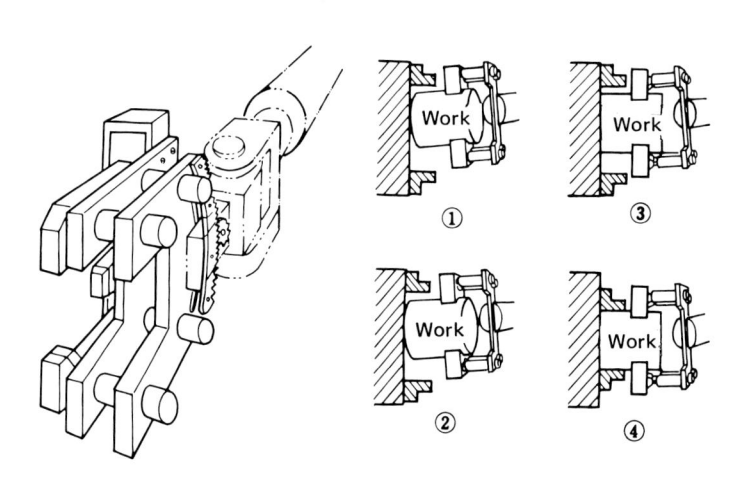

THREE FINGERS WITH AN AIR NOZZLE FOR CLEANING

Application No.	PAT. 1974-111356
Applicant	Yamatake Honeywell

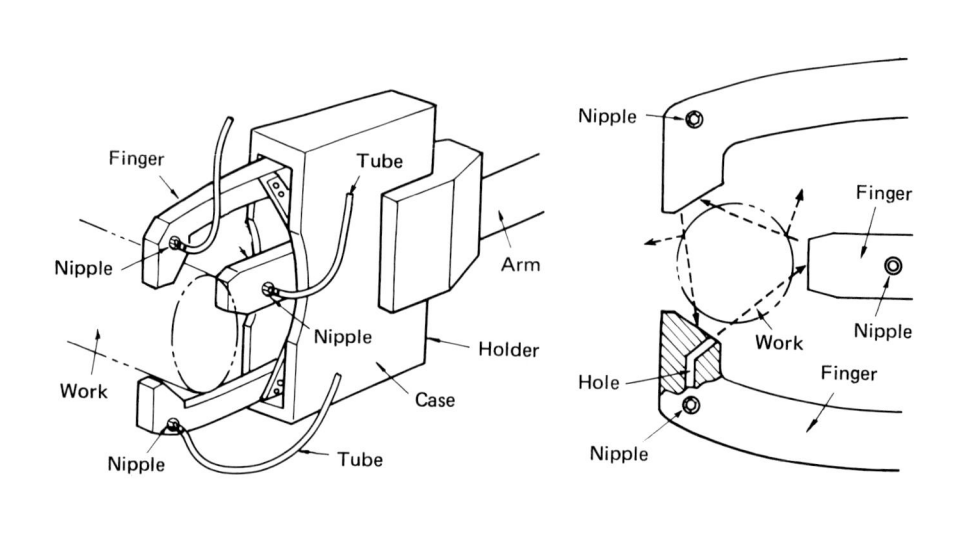

FINGERS FOR COMPOSITION WITH SENSOR FOR FITTING DIRECTION

Application No.	PAT. 1974-113366
Applicant	Hitachi

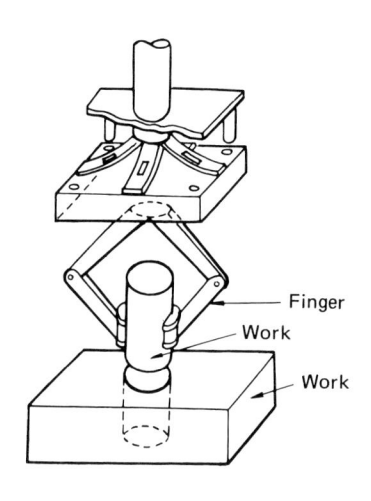

SLIDE HAND

Application No.	PAT. 1975-27265
Applicant	Tokico

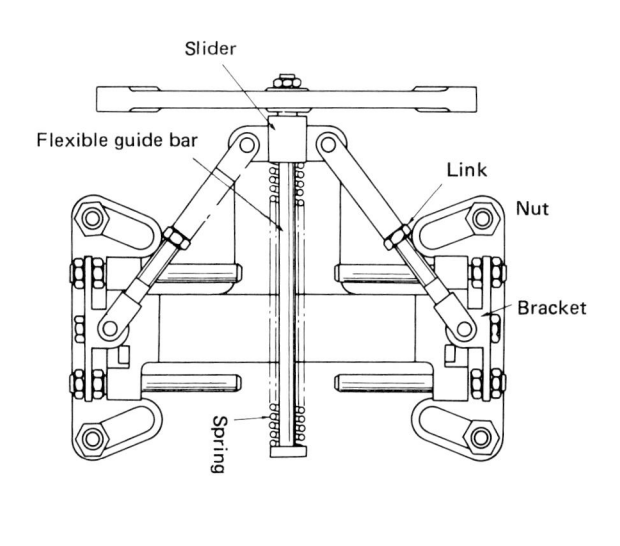

GRASPING HAND FOR LARGE-SCALE MATERIALS

Application No.	PAT. 1975-37156
Applicant	Hitachi

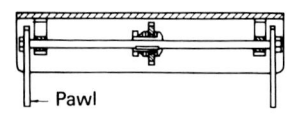

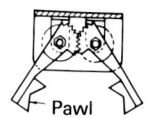

HAND FOR SETTING THE CENTER OF OBJECTS

Application No.	PAT. 1975-60970
Applicant	Tokyo Shibaura Elec.

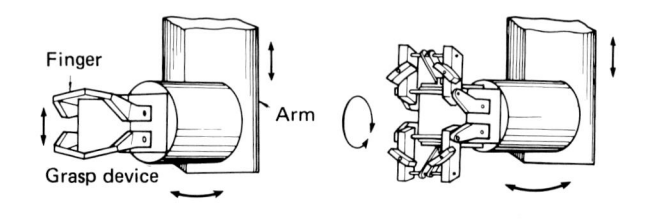

PARALLEL MOVABLE FINGER

Application No.	PAT. 1053627
Applicant	Daini Seikosha

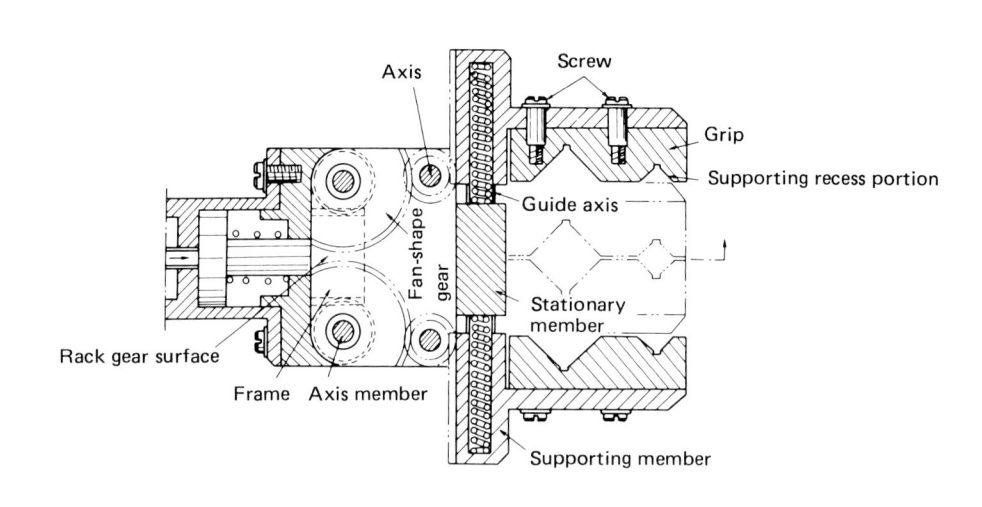

HAND WITH ADJUSTABLE PRESSURE

Application No.	PAT. 1975-131254
Applicant	Tokyo Shibaura Elec.

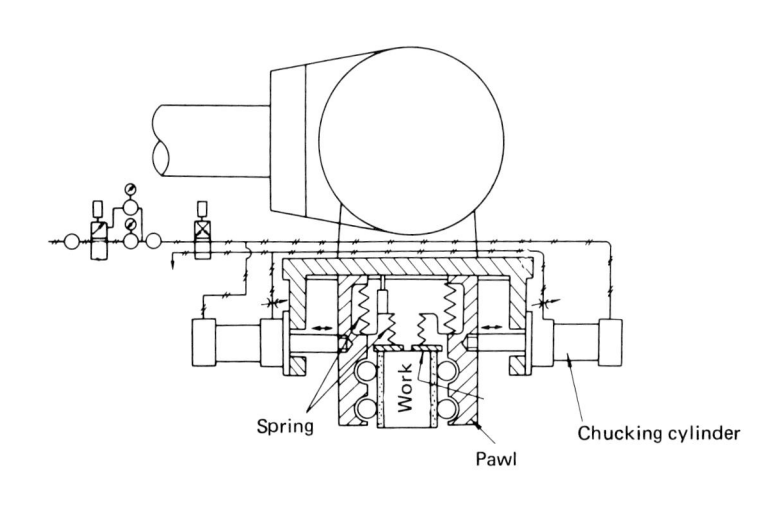

ROBOT HAND FOR COMPACT WORKING AREA

Application No.	PAT. 1975-160966
Applicant	Yamatake Honeywell

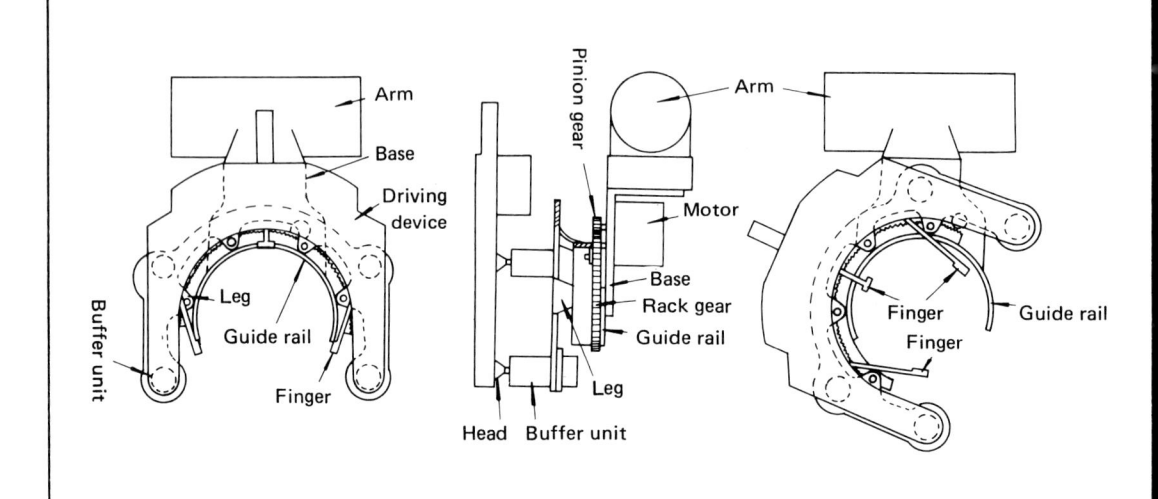

HAND WITH A CAM AND A SPRING

Application No.	PAT. 1976-23954
Applicant	Matsushita Elec.

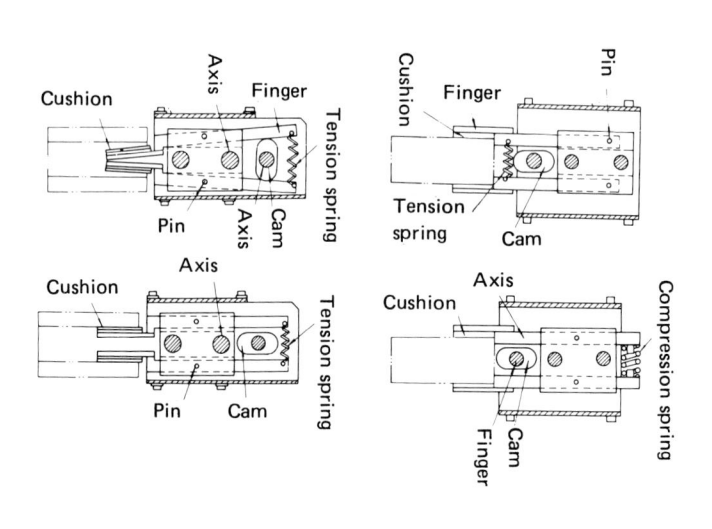

FINGERS FOR CYLINDRICAL OBJECTS

Application No.	1027040
Applicant	Fuji Electric

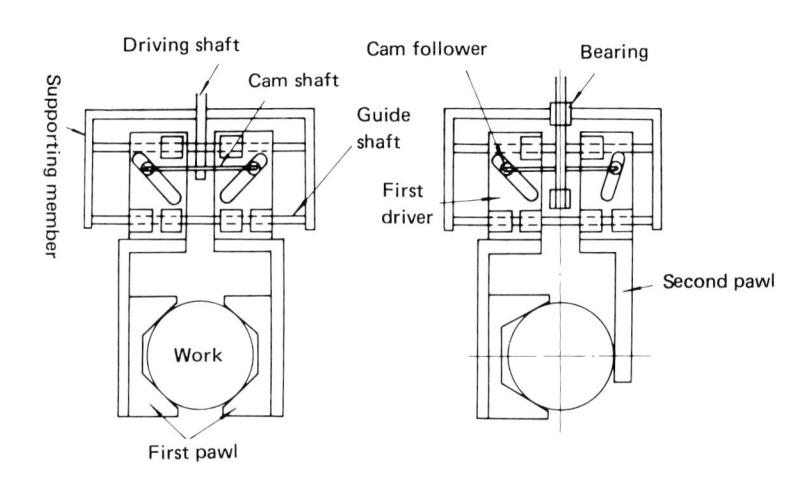

GRASPING DEVICE WITH A COUPLE OF HORIZONTALLY MOVABLE JAWS

Application No.	PAT. 1976-5764
Applicant	Hitachi

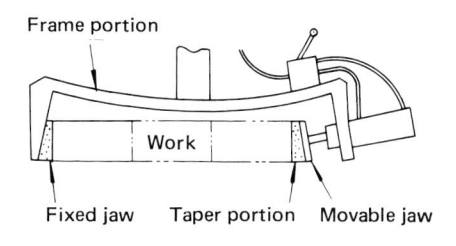

EASILY REMOVABLE HAND FOR VARIOUS CYLINDRICAL OBJECTS

Application No.	PAT. 1976-5770
Applicant	Kawasaki Heavy Industry

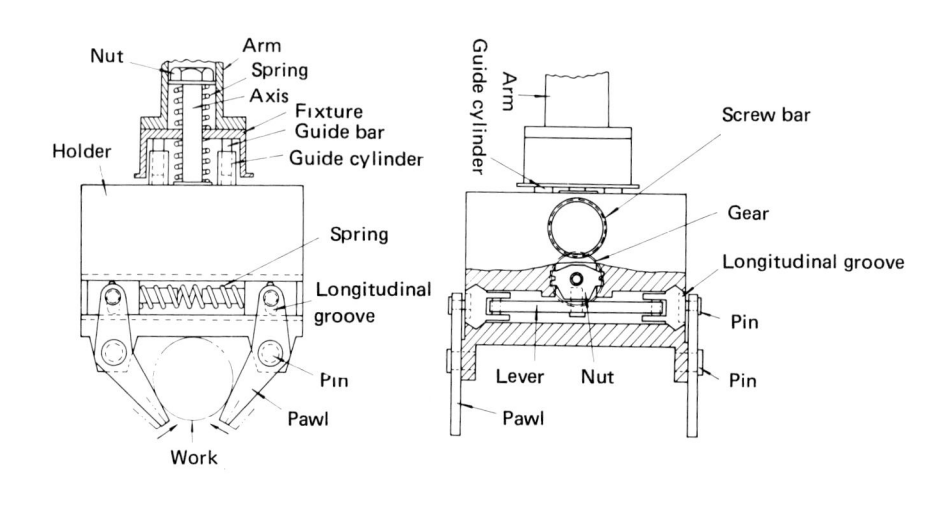

GRASPING DEVICE

Application No.	PAT. 1976-35949
Applicant	Motoda Electronics

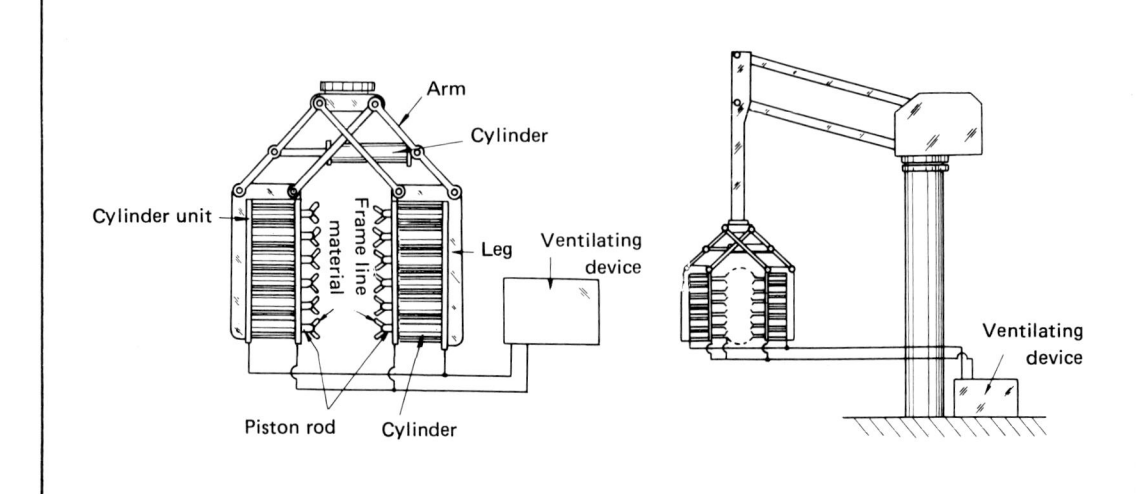

WORK GRASPING DEVICE

Application No.	PAT. 1976-39859
Applicant	Hitachi

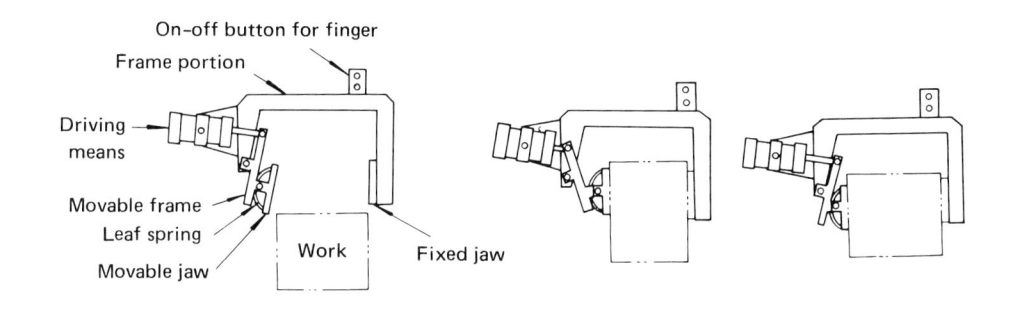

FINGERS WITH DIFFERENT CONCAVE PARTS

Application No.	PAT. 1976-112064
Applicant	Nippondenso

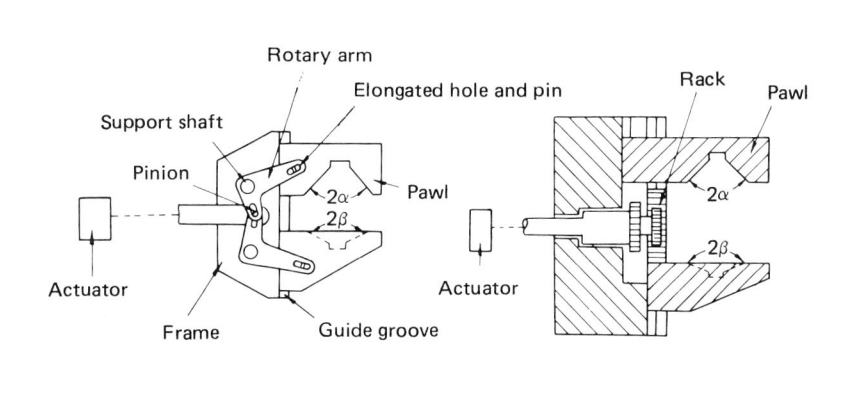

HAND WITH SMALL-DIAMETER GRASPING INSTRUMENT

Application No.	Utility model 1972-11267
Applicant	Toyoda Machine Works

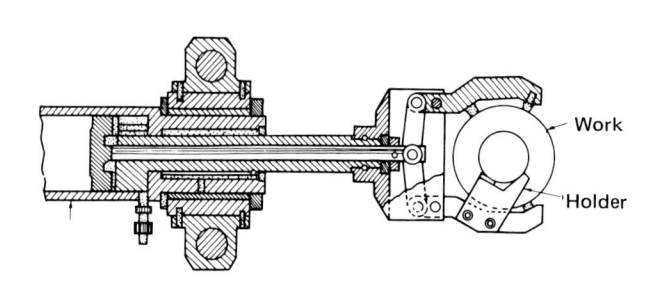

HAND FOR SHAFT-TYPE OBJECTS

Application No.	Utility model 1157768
Applicant	Aikoku Industry

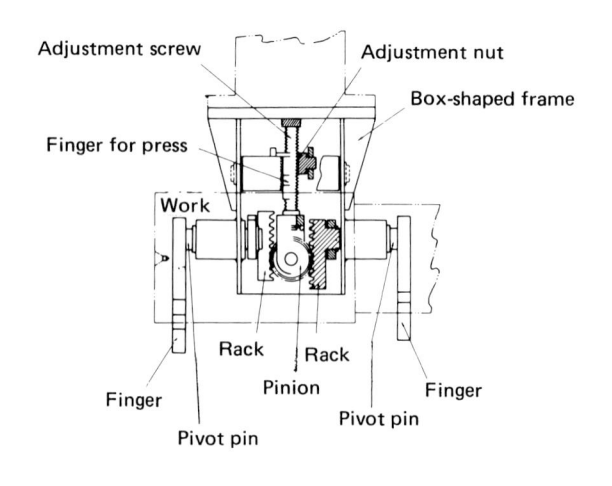

GRASPING DEVICE FOR MANIPULATORS

Application No.	Utility model 1974-68167
Applicant	Komatsu

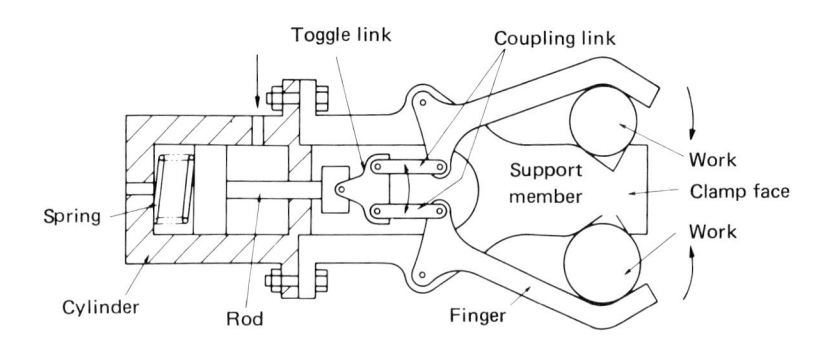

GRASPING DEVICE FOR VARIOUS OBJECTS

Application No.	Utility model 1974-141565
Applicant	Dowa Kogyo

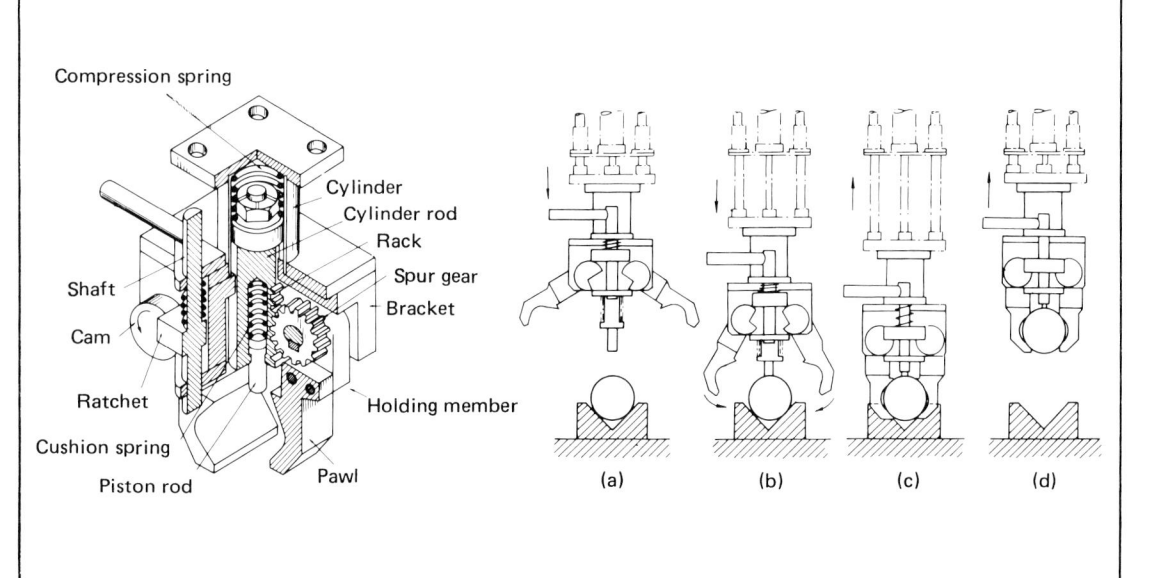

HAND WITH RACK AND PINION

Application No.	Utility model 1975-114076
Applicant	Hitachi Seiki

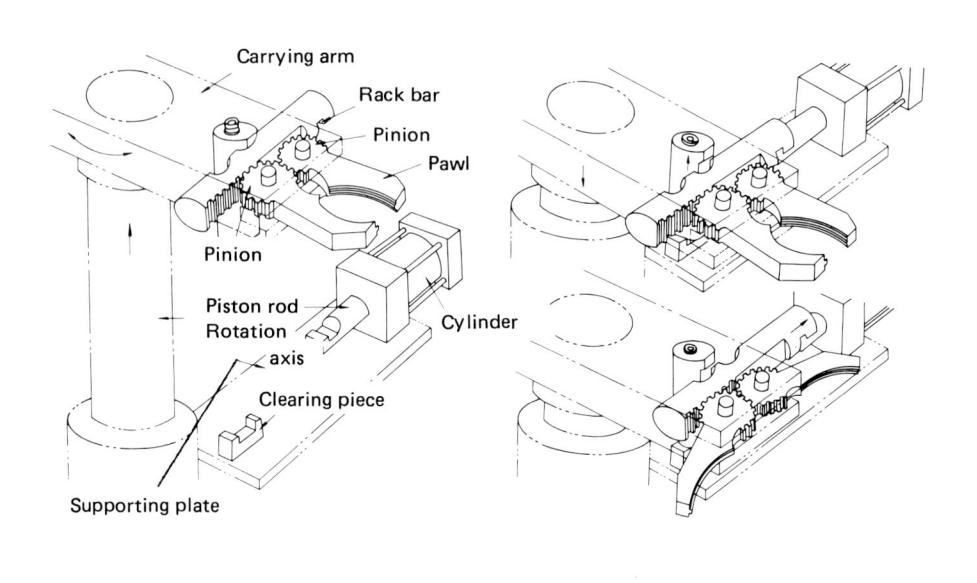

ROTATIONAL GRASP HAND

Application No.	Utility model 1975-114077
Applicant	Tokico

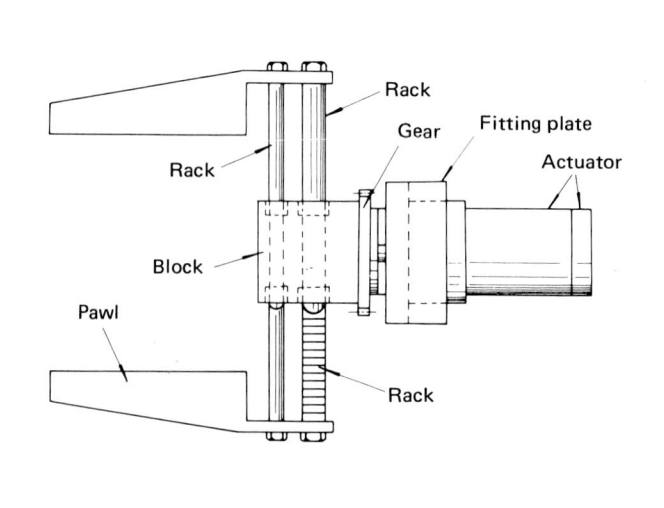

HAND WITH NORMAL SCREW AND REVERSING SCREW

Application No.	Utility model 1975-130069
Applicant	Howa Sangyo

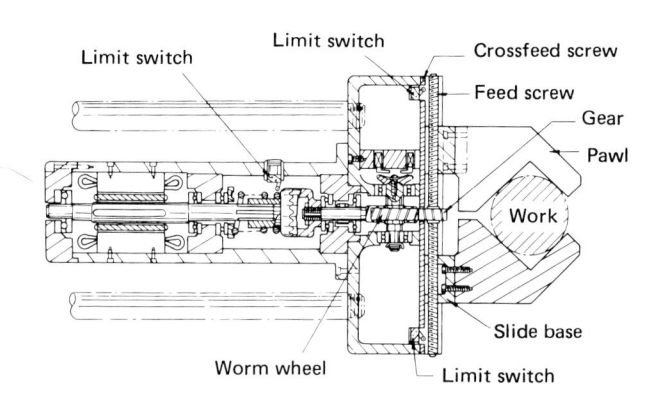

GRASPING HAND WITH MAGNETS

Application No.	Utility model 1975-1373403
Applicant	Kogyosha

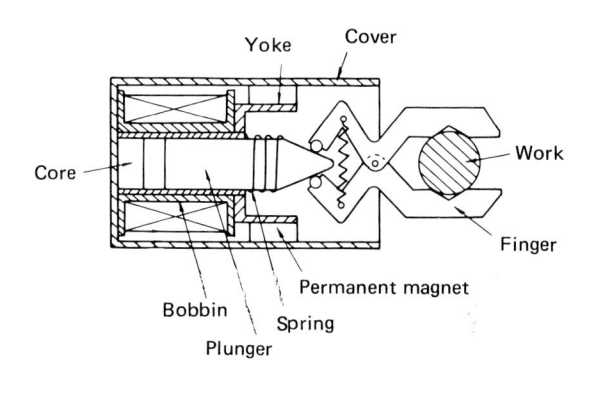

REMOTE-CONTROL GRASPING INSTRUMENT

Application No.	Utility model 1975-141383
Applicant	Japan Atomic Energy Research Institute

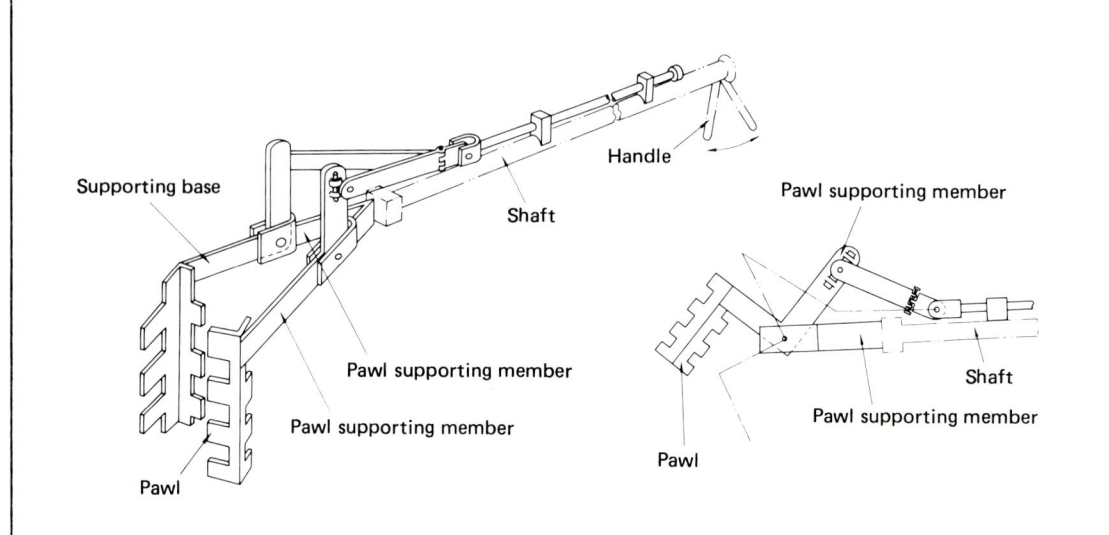

HAND POWERED BY THE WEIGHT OF OBJECTS

Application No.	Utility model 1975-143284
Applicant	Hitachi

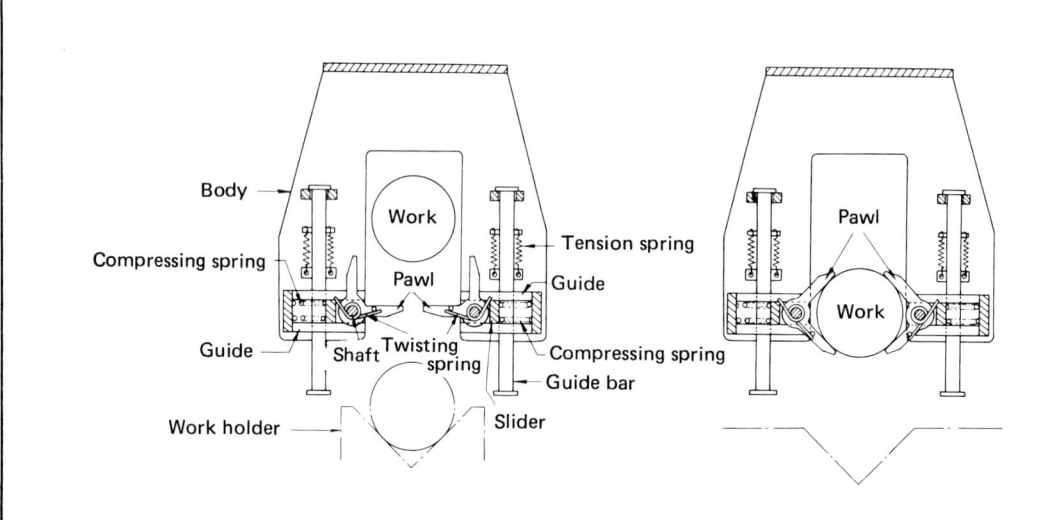

FREE TYPE: SNAP

HAND FOR EXCHANGE OF TOOLS

Application No.	Utility model 1975-144374
Applicant	Hitachi Seiki

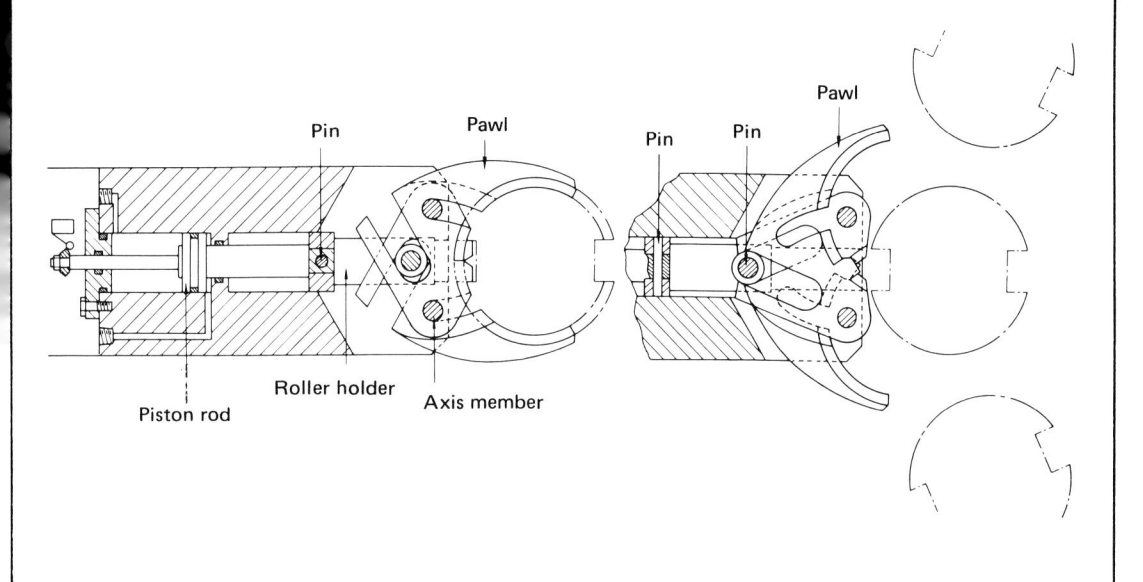

HAND WITH CHUCKS FOR CYLINDRICAL OBJECTS

Application No.	Utility model 1976-37377
Applicant	Honda Engineering

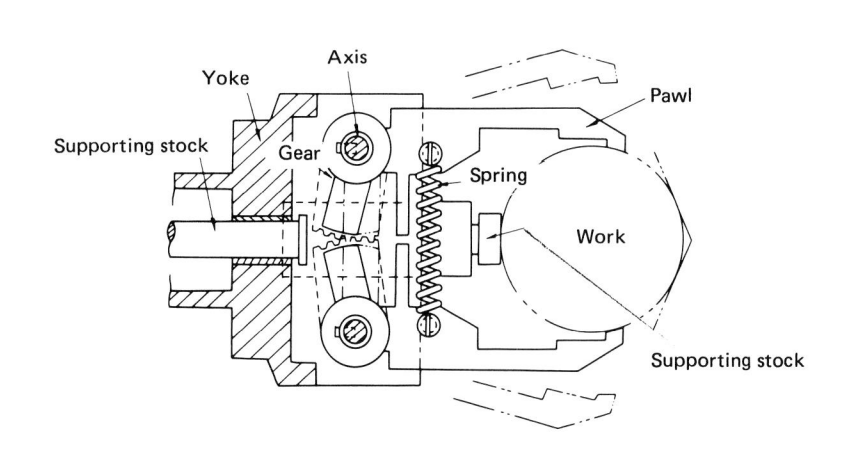

HAND WITH ELECTRIC REVOLUTION MECHANISM

Application No.	Utility model 1976-63377
Applicant	Tokyo Shibaura Elec.

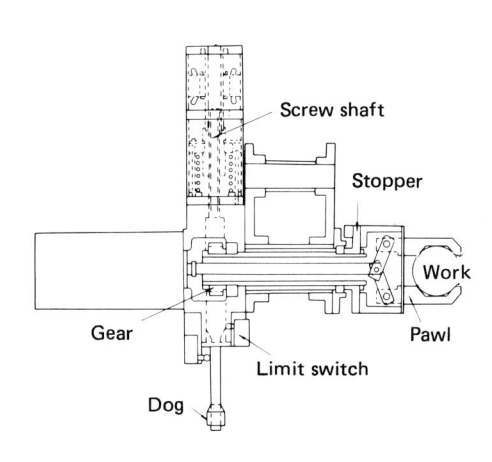

HAND WITH A FIXED PAWL AND A MOVABLE PAWL

Application No.	Utility model 1976-68079
Applicant	Kondo Seisakusho

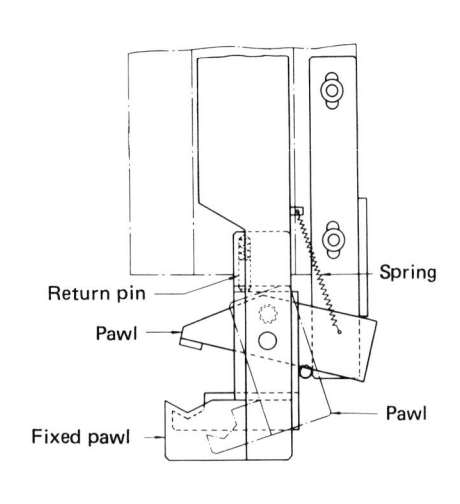

ELECTRIC GRASP HAND

Application No.	
Applicant	

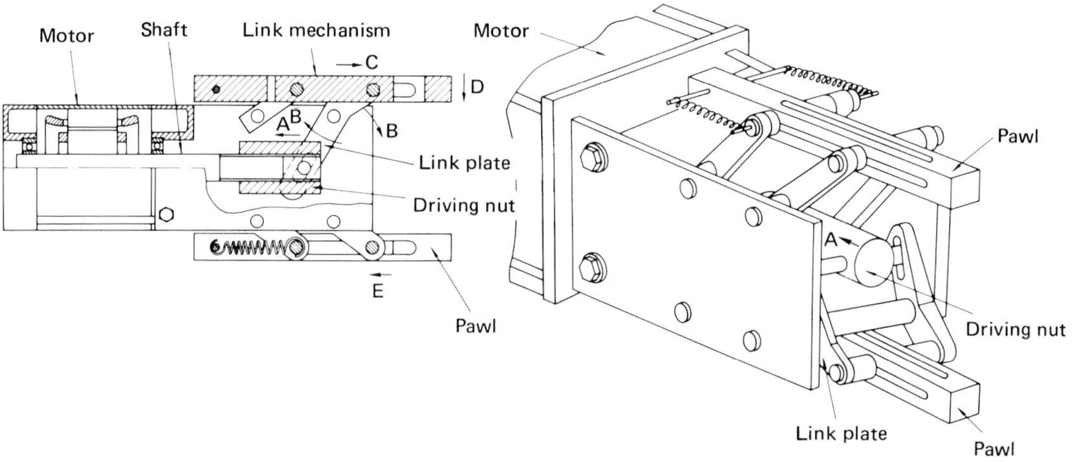
Motor
Shaft
Link mechanism
C
D
A
B
B
Link plate
Driving nut
E
Pawl
Motor
Pawl
A
Driving nut
Link plate
Pawl

Application No.	
Applicant	

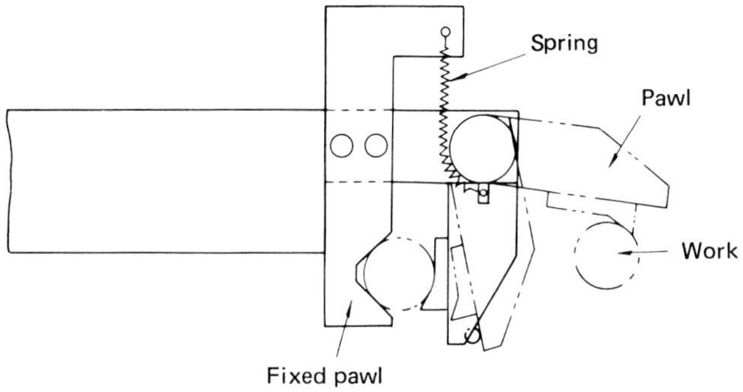
Spring
Pawl
Work
Fixed pawl

HAND WITH A FREE ROTATIONAL PAWL

Application No.	Utility model 1976-132773
Applicant	Hitachi

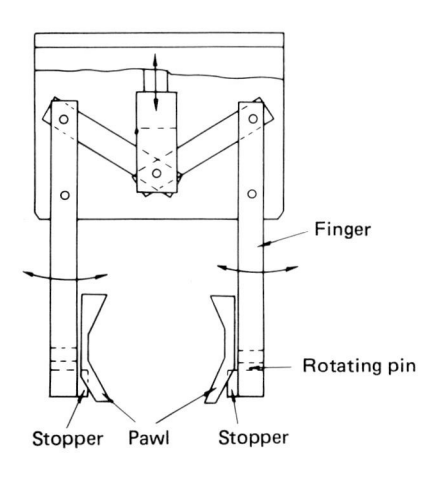

GRASPING HAND FOR BOTH INNER DIAMETER AND OUTER DIAMETER

Application No.	Utility model 1976-132774
Applicant	Yamatake Honeywell

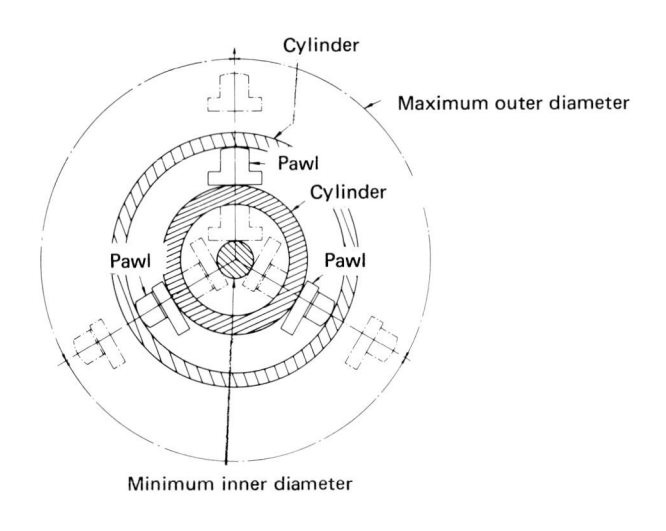

HAND WITH TWO CYLINDERS

Application No.	Utility model 1976-145180
Applicant	Kondo Seisakusho

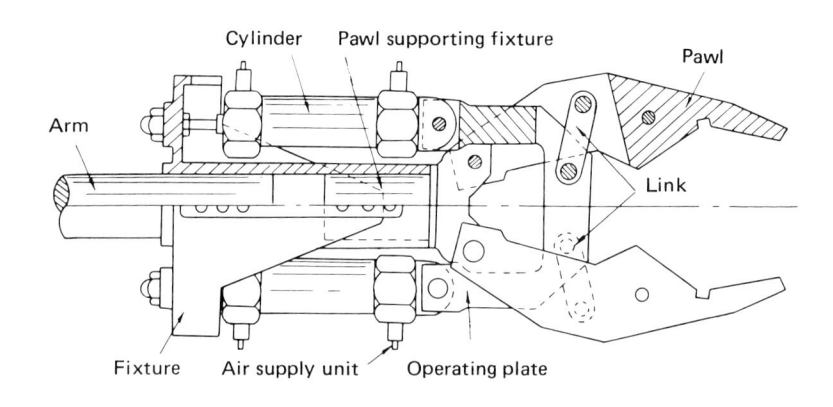

HAND WITH A RECTILINEAR PAWL AND A ROTATIONAL PAWL

Application No.	Utility model 1976-148174
Applicant	Hitachi

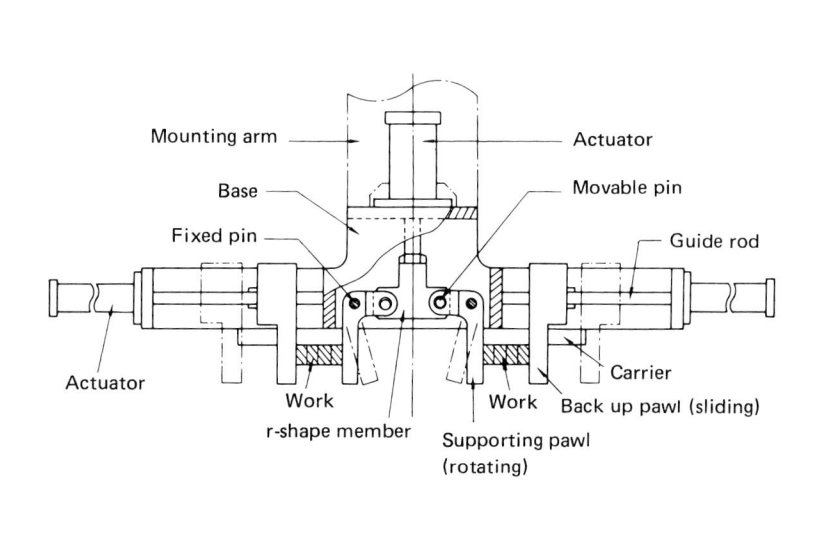

FINGERS FOR MODIFYING THE DISAGREEMENT OF CENTERS OF DIFFERENT RODS

Application No.	PAT. 1973-9501
Applicant	Yaskawa

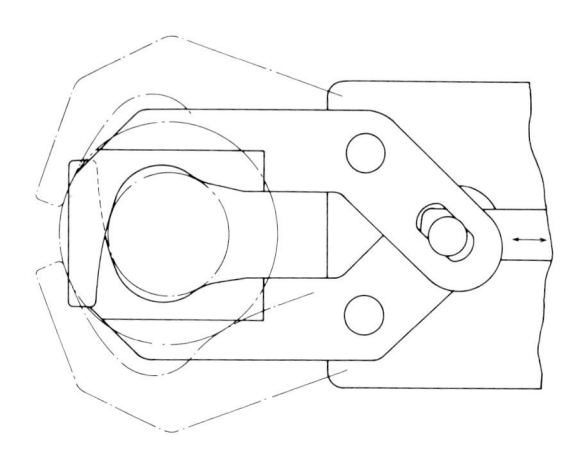

FINGERS OF FLEXIBLE MATERIAL

Application No.	PAT. 1974-36306
Applicant	Asahi Kiko

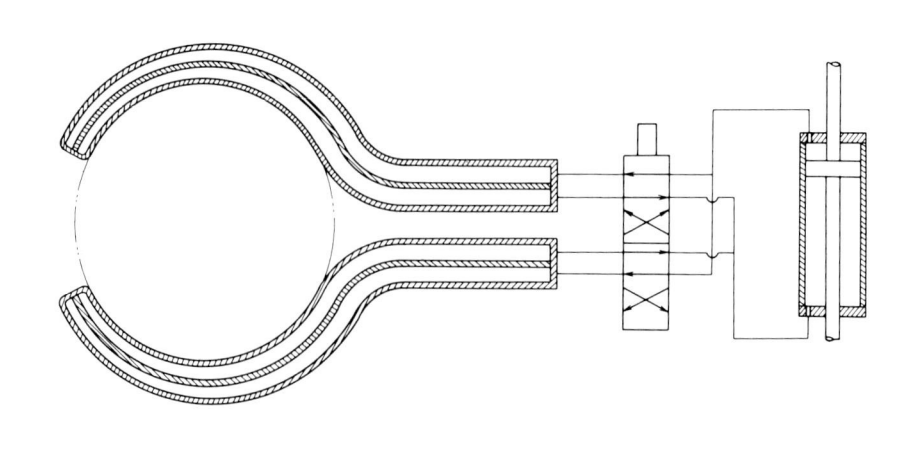

HAND WITH A SELF-LOCK MECHANISM

| Application No. | PAT. 1974-64155 |
| Applicant | Fujitsu |

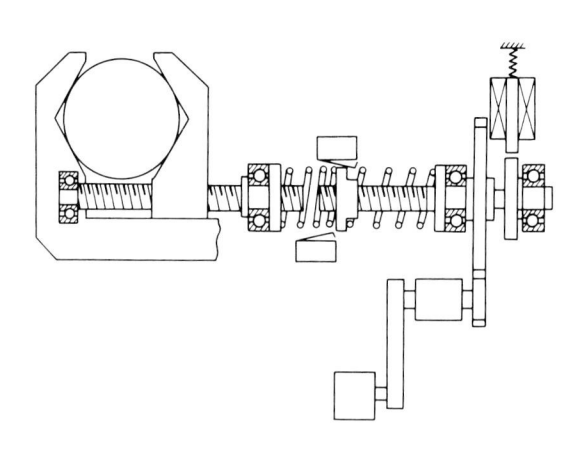

PUSH TOUCH HAND WITH AN OBLIQUE GUIDE

| Application No. | PAT. 1974-64156 |
| Applicant | Fujitsu |

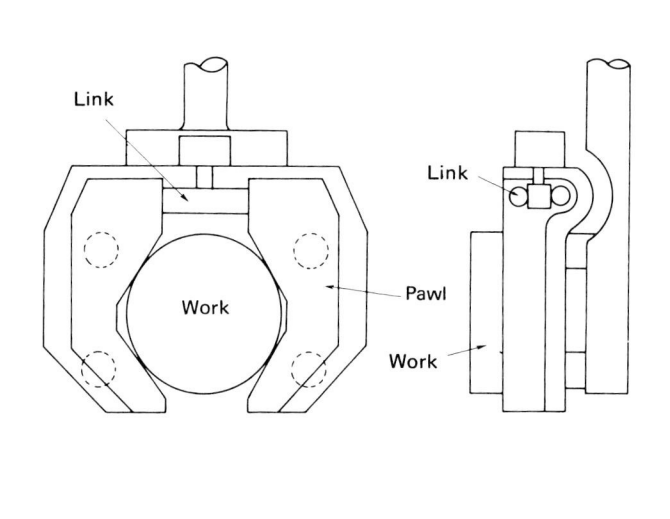

FINGERS FOR OBJECTS OF VARIOUS SCALES

Application No.	PAT. 1974-64157
Applicant	Fujitsu

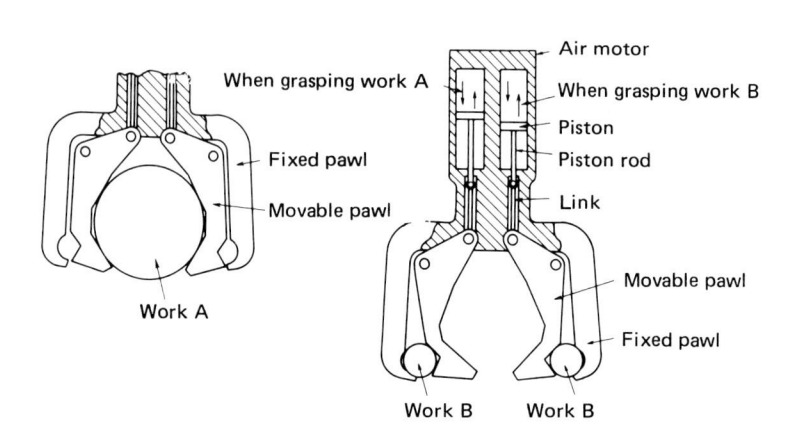

When grasping work A — Fixed pawl, Movable pawl, Work A

When grasping work B — Air motor, Piston, Piston rod, Link, Movable pawl, Fixed pawl, Work B, Work B

HAND FOR INSERTING PIPES

Application No.	PAT. 1974-82053
Applicant	Unic

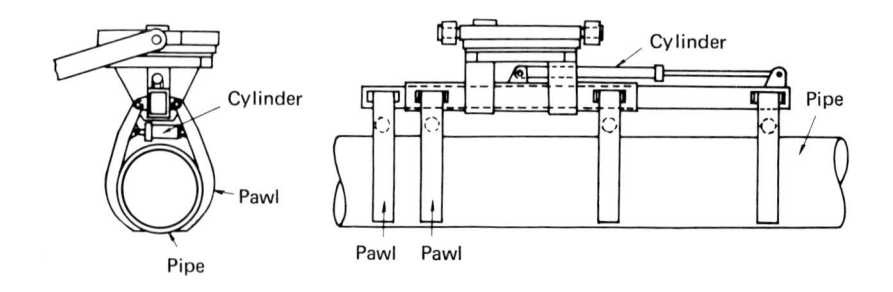

Cylinder, Pawl, Pipe

Cylinder, Pawl, Pawl, Pipe

CENTERING FINGERS OF GUIDE BODY

Application No.	PAT. 1974-118156
Applicant	Hitachi

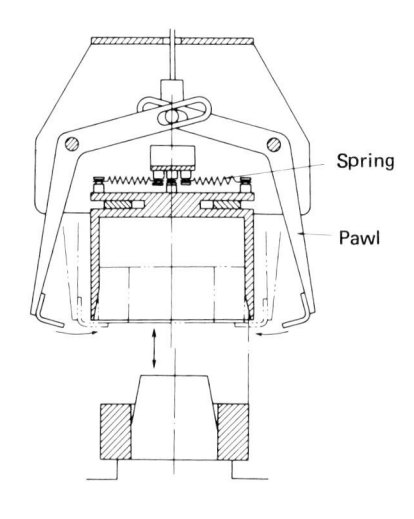

FINGERS HOLDING THE CENTER OF RODS WITH DIFFERENT DIAMETERS

Application No.	PAT. 1974-127360
Applicant	Tokyo Shibaura Elec.

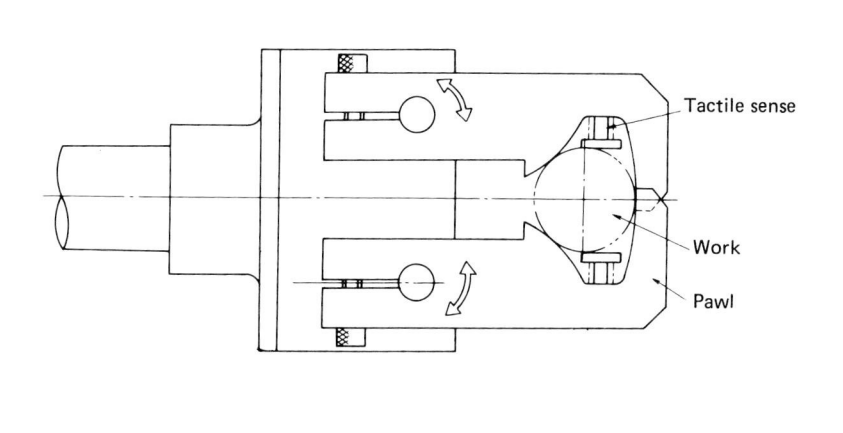

FINGERS WITH A RELIEF MECHANISM

Application No.	PAT. 1975-8255
Applicant	Hitachi

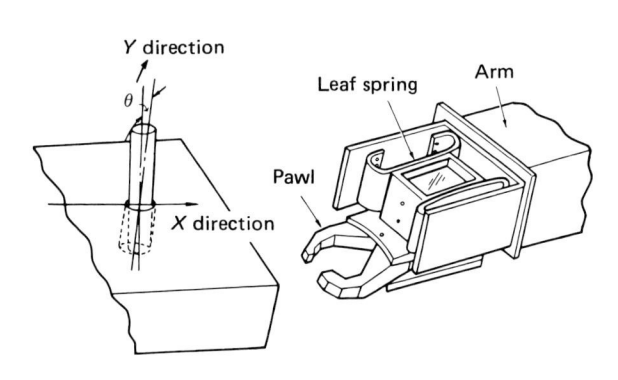

HAND SWITCHED BY PRESSURE OF TOUCHING THE GROUND

Application No.	PAT. 1975-58754
Applicant	Koga Kogyo

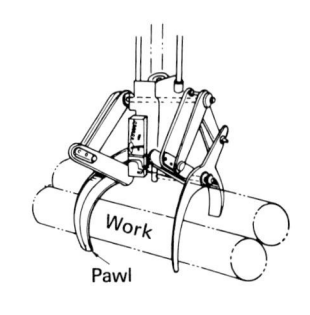

FREE TYPE: GRIP

HAND FOR EASY AND RELIABLE GRASPING OF LARGE-SCALE MATERIALS

Application No.	PAT. 1975-147061
Applicant	Kitai Ironworks

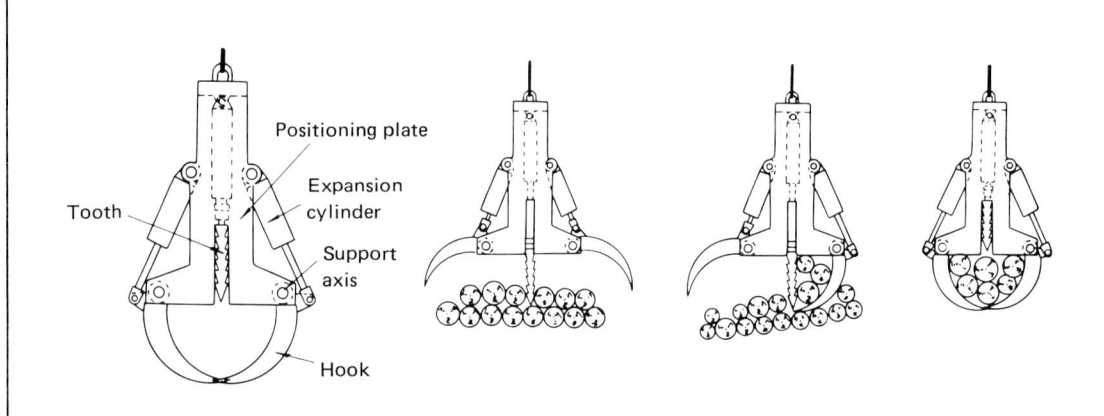

CLAMP AND PICK-UP DEVICE

Application No.	Utility model 1974-27872
Applicant	Kubota Ironworks

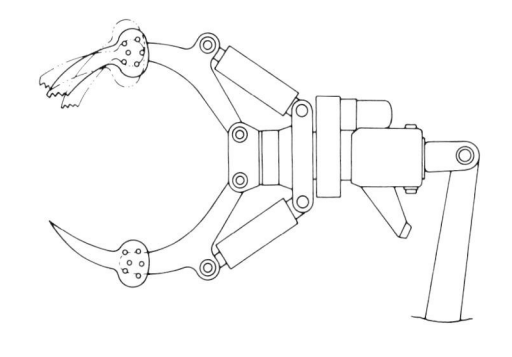

SUCTION HAND MOVABLE FORWARD AND BACKWARD

Application No.	PAT. 1975-135763
Applicant	Shiroyama Kogyo

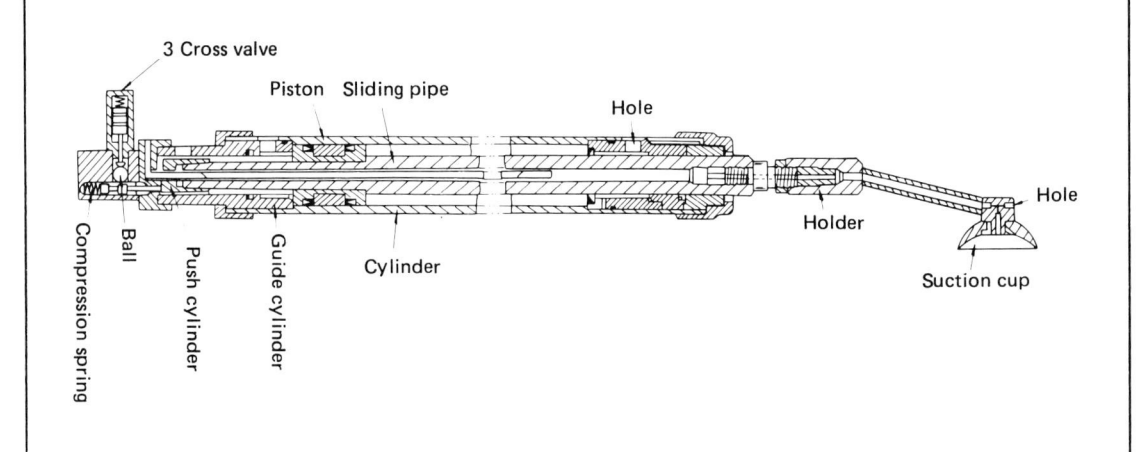

SUCTION HAND WITH READY KNOCKOUT PIN

Application No.	Utility model 1289735
Applicant	Daini Seikosha

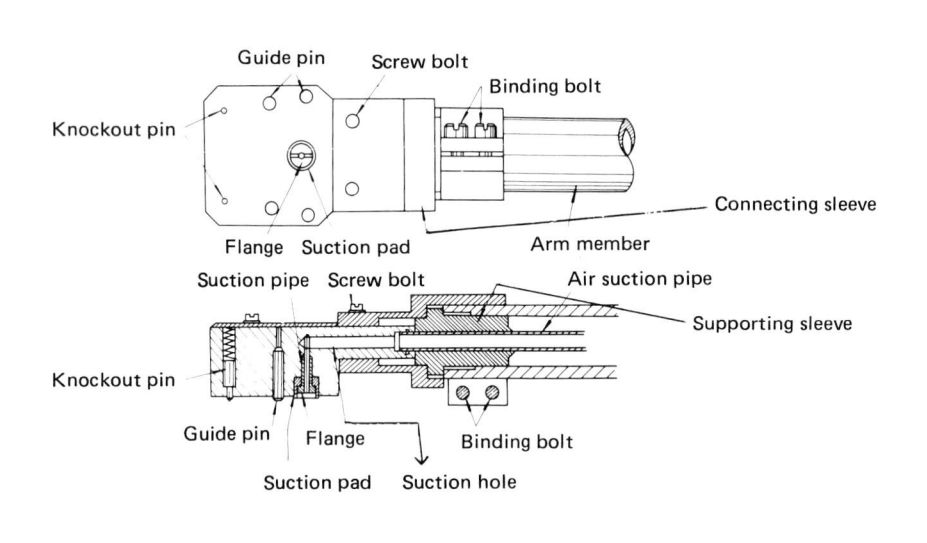

SUCTION AND PULL HAND

Application No.	Utility model 1975-117576
Applicant	Nissan Motor

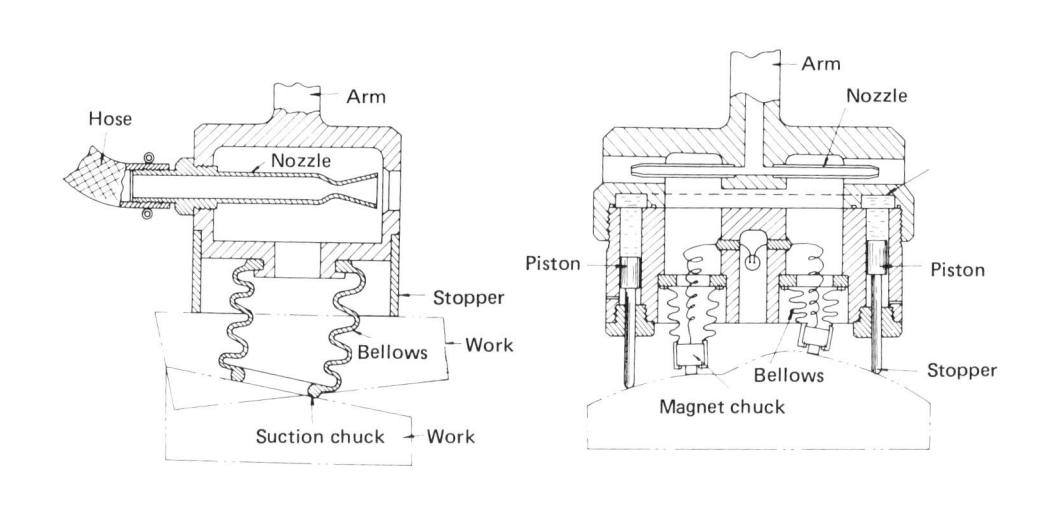

SUCTION HAND WITH SPOOL CONTAINING PRESSURE ROOM

Application No.	Utility model 1366048
Applicant	Toyota Motor

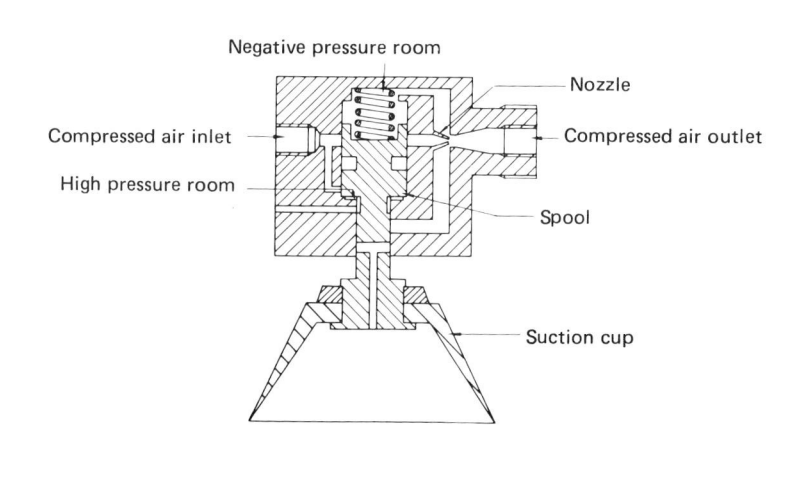

PULLING FINGER WITH MAGNETS

Application No.	PAT. 1975-69755
Applicant	Ichiko Kogyo

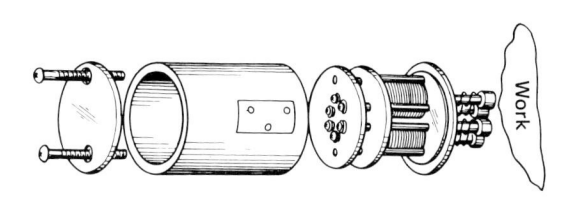

MAGNETIC HAND WITH MULTIPLE ABSORPTION DISKS

Application No.	Utility model 1976-126570
Applicant	Kayaba Industry

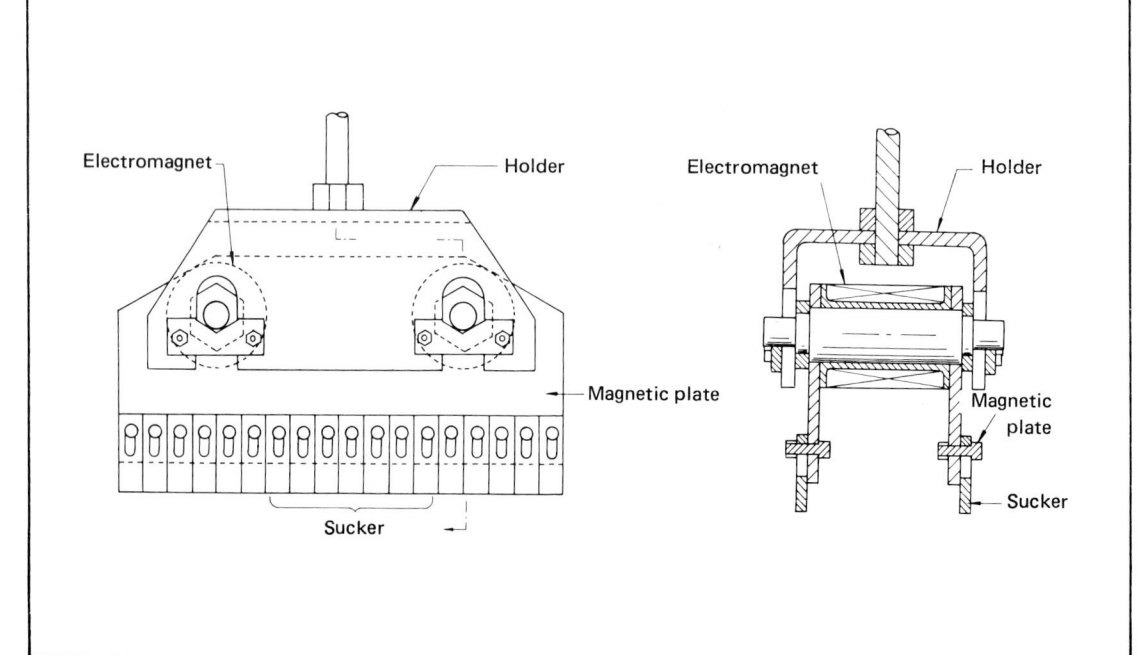

OTHER TYPE

HAND STRETCHING FROM INNER SIDE

Application No.	PAT. 9452535
Applicant	Aikoku Industry

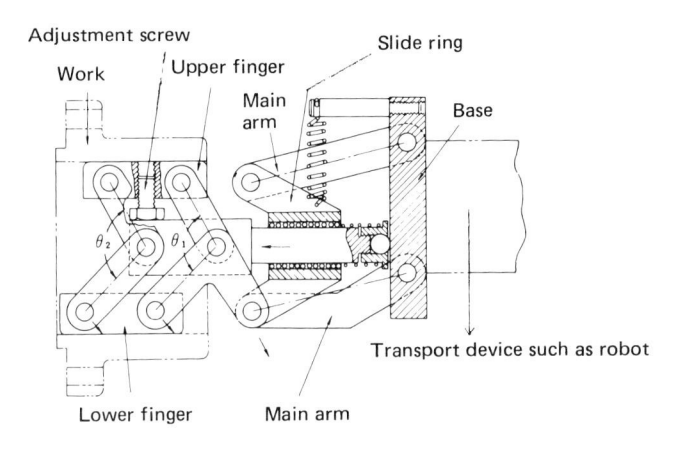

FINGERS WITH PRESSURE WELDING MATERIALS

Application No.	PAT. 1974-87060
Applicant	Mitsubishi Electric

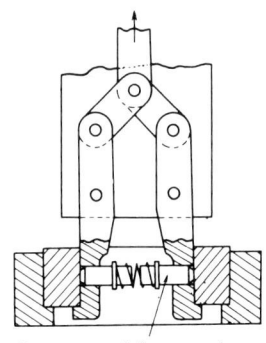

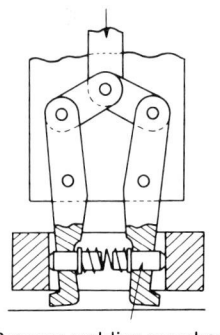

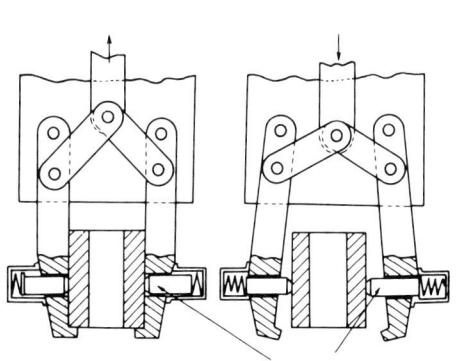

HAND INDIFFERENT TO SLIGHT ATTITUDE CHANGE

Application No.	PAT. 1975-32659
Applicant	Fanuc

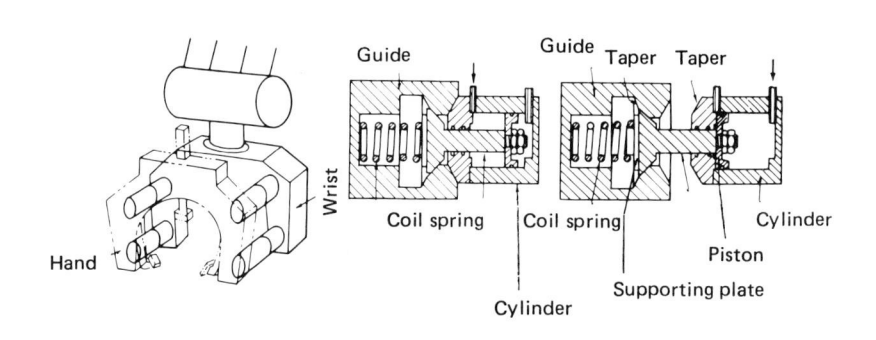

SPOT WELDING HAND WITH DIRECTION CONTROL

Application No.	PAT. 1975-72374
Applicant	

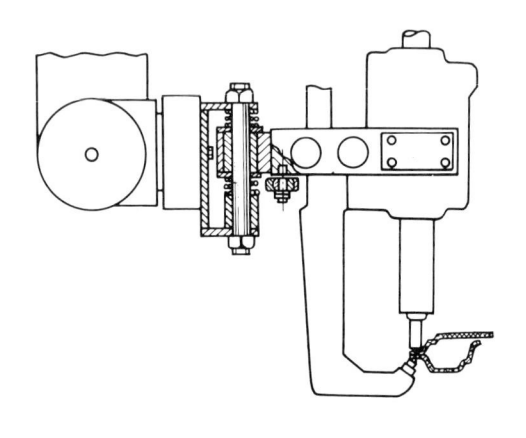

OTHER TYPE

HAND FOR TAKING OUT CYLINDRICAL MATERIALS

Application No.	PAT. 1976-43561
Applicant	

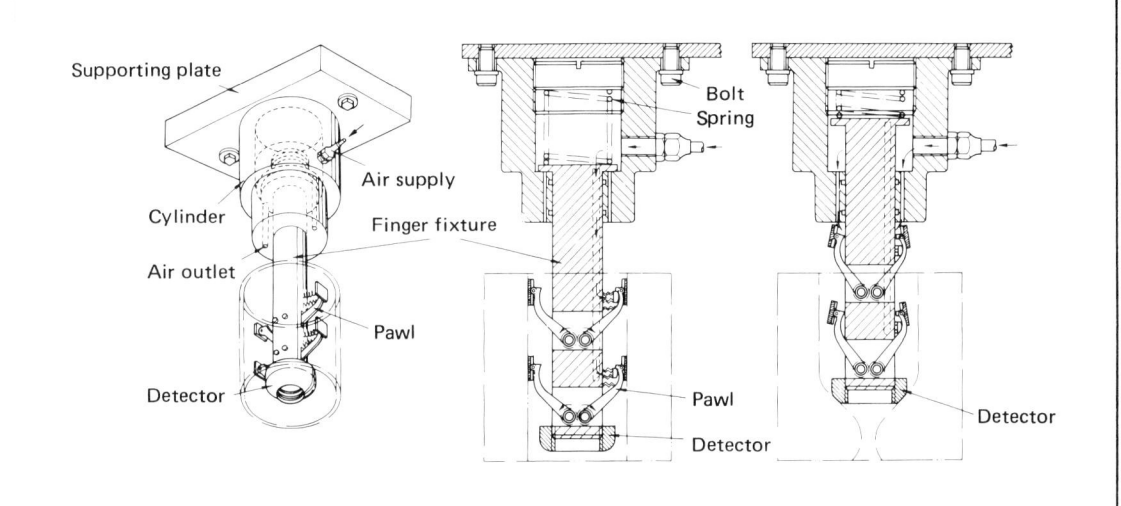

UNIVERSAL HAND FOR DOUGHNUT-LIKE OBJECTS

Application No.	

GRASPING HAND FOR THE INNER STAGE OF WORK

Application No.	PAT. 1976-50169
Applicant	Hitachi

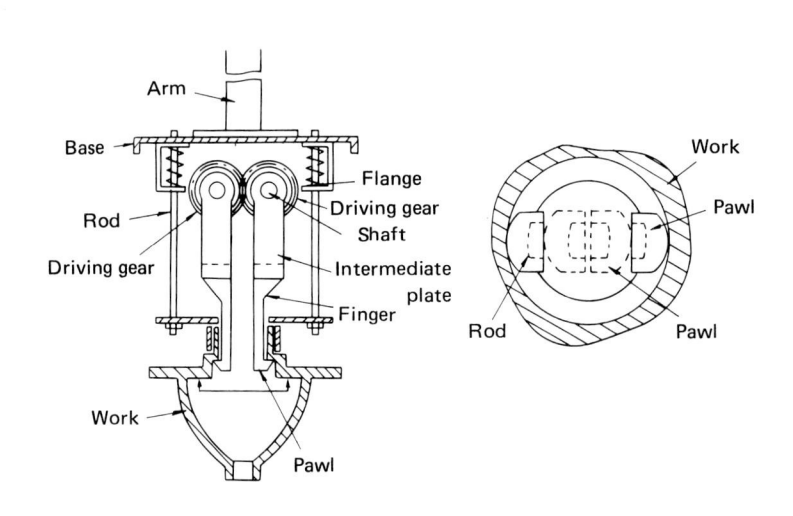

HAND WITH ADJUSTABLE GRIPPING CENTER

Application No.	PAT. 1976-55563
Applicant	Hitachi

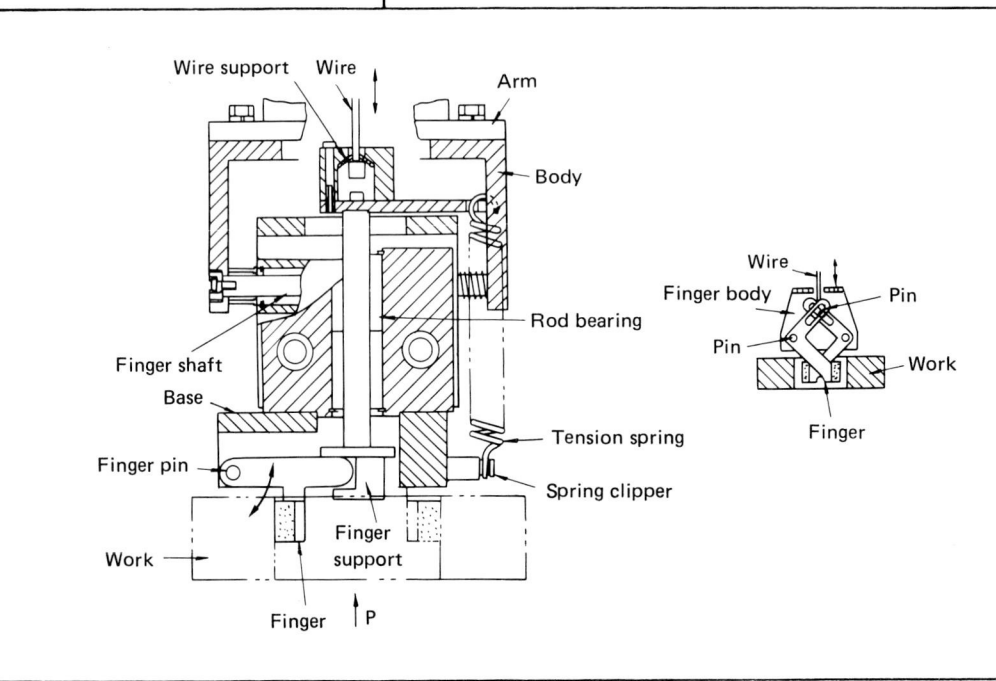

OTHER TYPE

PULLOUT HAND FOR HOUSE BOXES

Application No.	Utility model 1974-115865
Applicant	Ishikawajima-Harima

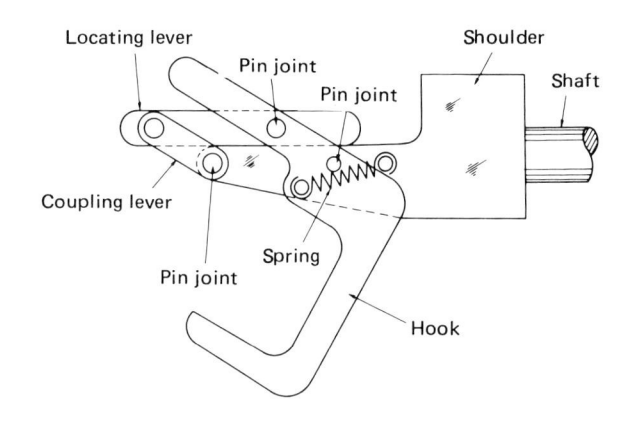

GRASPING HAND CONSISTING OF MULTIPLE JOINTS

Application No.	Utility model 1976-89674
Applicant	Yasukawa

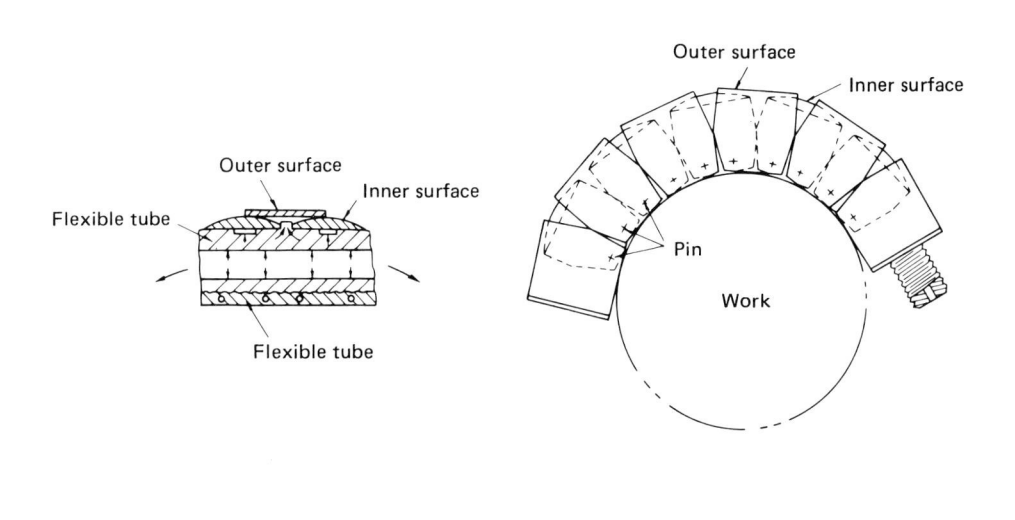

HAND WITH ROTATION-TUNABLE COUPLE OF FINGERS

Application No.	Utility model 1976-121978
Applicant	

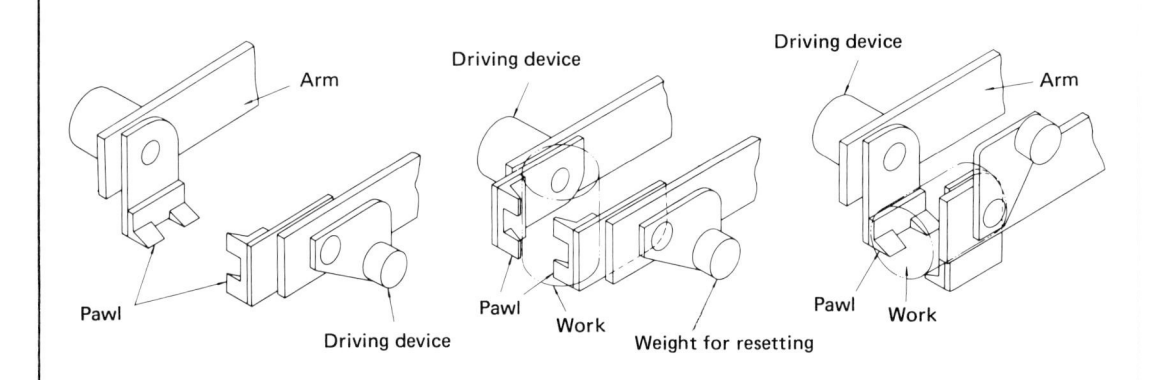

HAND WITH A MACHINE MECHANISM IN THE MAIN

Application No.	Utility model 1976-143976
Applicant	Tokyo Shibaura Electric

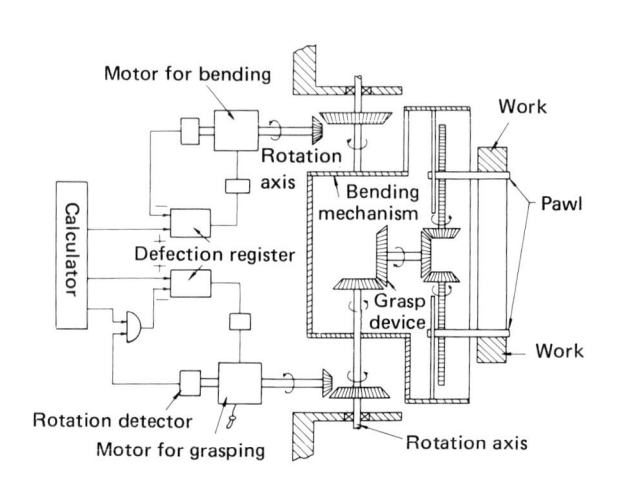

Appendix

Actual Mechanical Hands

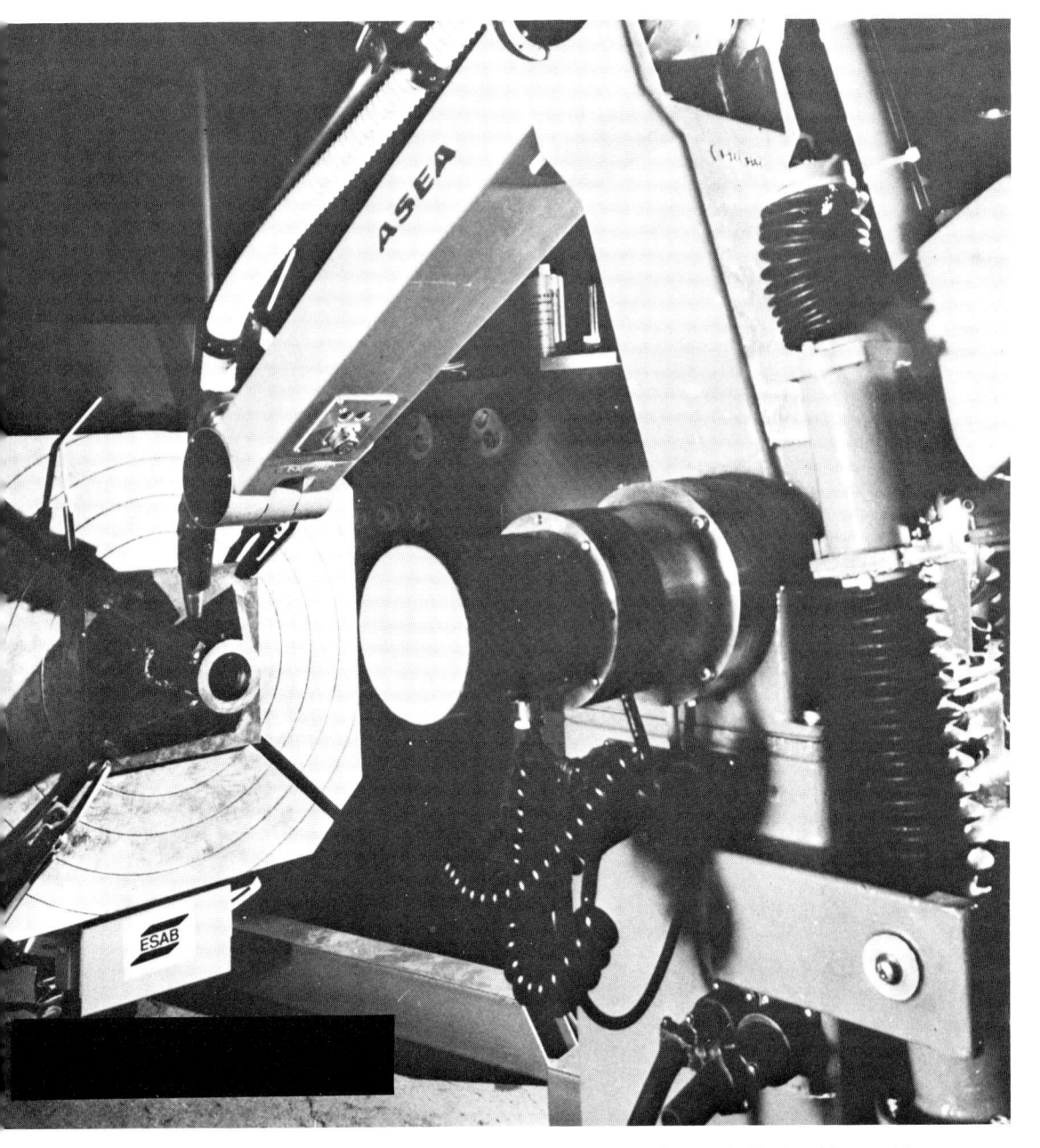

Figure A.1 Welding being done by an industrial robot (made by ASEA Corp., Sweden). The hand is a welding gun.

Figure A.2 Mechanical hand powered by a motor grasping work of complex shape.

Figure A.3 A hand powered by a motor and grasping cylindrical body contributes to the automation of conveyor systems.

Figure A.4 Mechanical hand transferring a TV Brown tube. It is capable of handling fragile goods well.

Figure A.5 Hand loading or unloading in the work process of rotors.

Figure A.6 Hand used in high-frequency hardening of gears. It is strong enough to withstand a harsh environment.

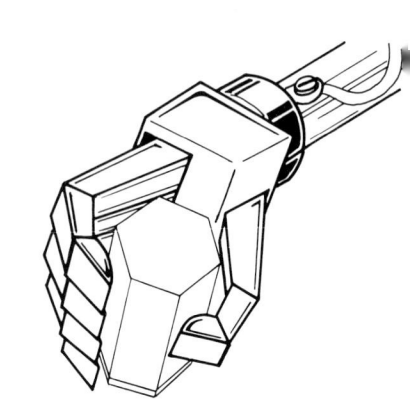

Figure A.7 Hand exchanging the jigs in shaft work.

Figure A.8 Hand setting a work on a press machine. It plays an important role in dangerous work.

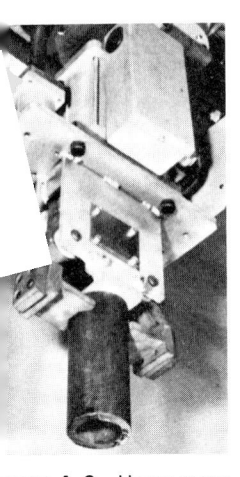

Figure A.9 Hang-type hand grasping a cylinder.

Figure A.10 Hand grasping a shaft at two positions.

Figure A.11 Hand picking up two brackets for cars at the same time. The hand has a locating mechanism.

Figure A.12 Hand transporting wound coils; it can handle three simultaneously.

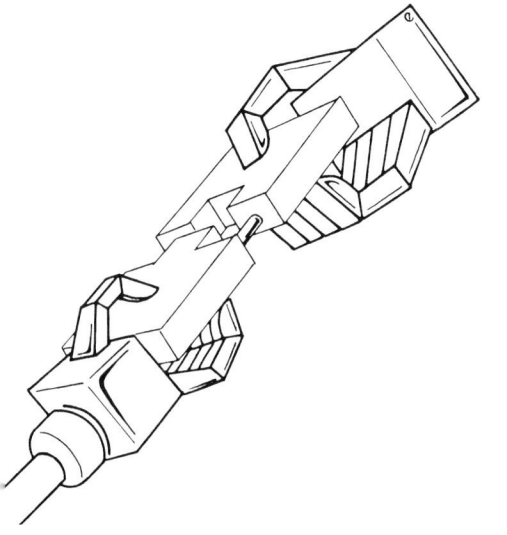

Figure A.13 Robot equipped with a grasp wrist for drills in grinding a drill. The wrist has the relief mechanism in the stretch direction.

Figure A.14 Vacuum hand transferring a washing machine.

Figure A.15 Hand pulling out a formed body from emergency forming machine and about to transfer it to the next process. It is powered by oil pressure.

Figure A.16 Absorption hand handling an electronic range.

Figure A.17 Absorption hand transferring a stereo speaker box.

Figure A.18 Mechanical hand transporting a block, powered by a motor.

Figure A.19 Hand searching the position of a body, equipped with an air sensor.

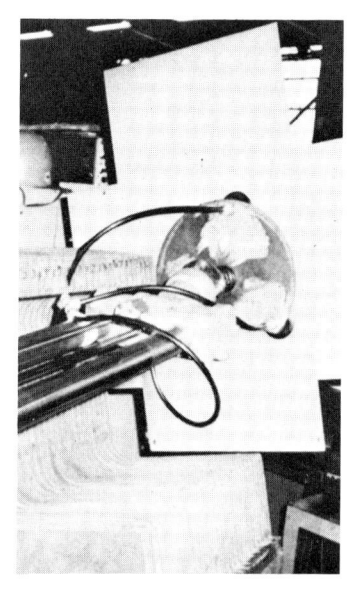

Figure A.20 Vacuum gripper taking out a blank from the magazine and transferring it to a press die.

Figure A.21 Vacuum hand handling goods made of glass.

213

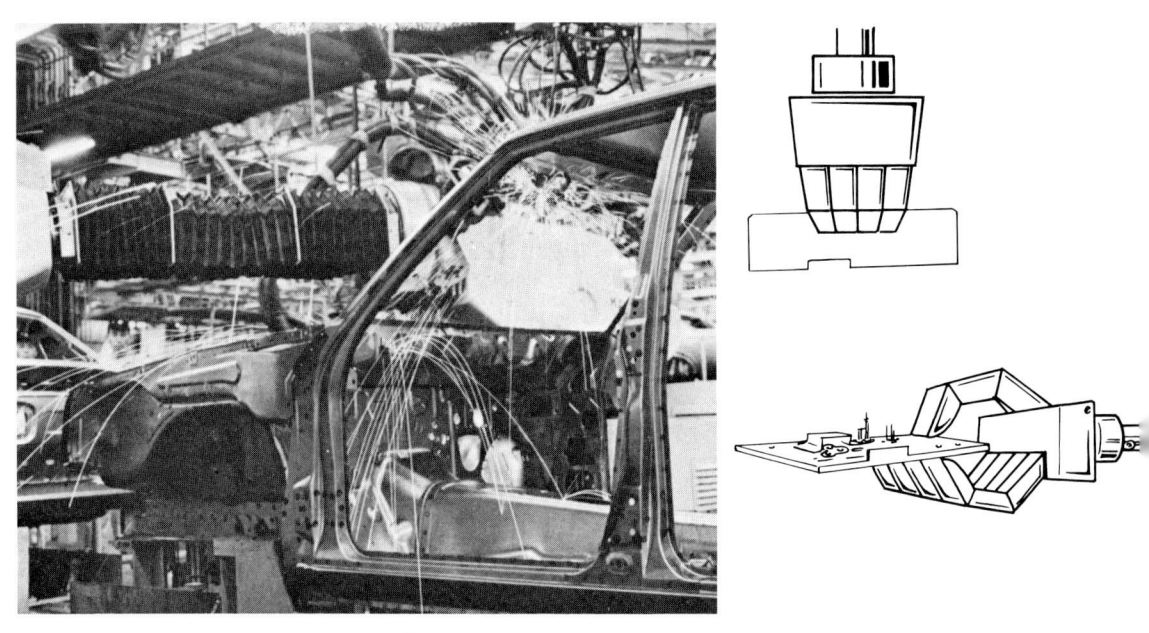

Figure A.22 Industrial robot doing spot welding in an automobile factory. It achieves high speed and exact welding.

Figure A.23 Industrial robot used in a wax painting process. It releases humans from dirty and foul-smelling work.

Index

Manufacturers and Power Sources

Manufacturer	Power Source			Manufacturer	Power Source		
	Electric Power	Hydraulic Pressure	Air Pressure		Electric Power	Hydraulic Pressure	Air Pressure
Chuo University	37-a 76-c		37-a	Kayaba Industry Co., Ltd.	86-c 140-f	68-b 69-b 70-b 86-c 140-f	
Daikin Kogyo Co., Ltd.	137-f	94-c 95-c 114-d 138-f	99-c 144-g	Keiaisha Mfg. Co., Ltd.			131-e 132-e 141-f
Dendensha Seisakusho		149-g	148-g	Keiaisha Seisakusho			103-c 104-c
Electrotechnical Laboratory	38-a 80-c			Kumamoto University	49-a		
FIT Co.	129-e 130-e		120-d 129-e 130-e	Kyobo			117-d 118-d 119-d
Kawasaki Heavy Industries, Ltd.	147-g		59-b 60-b 61-b 97-c 127-e 147-g	Kyoto University	56-a		
				Kyushu University	85-c	85-c	

Manufacturer	Electric Power	Hydraulic Pressure	Air Pressure
Marine Science and Technology Center		87-c	
Marol Company, Ltd.	89-c 90-c	71-b 72-b 89-c 90-c 91-c 92-c 93-c 115-d 146-g	
Mechanical Engineering Laboratory	46-a 79-c 84-c	54-a 84-c 88-c	
Ministry of Labor Research on Industry and Safety	121-d (Also: viscous fluid, 121-d)		
Nakamura-kiki Engineering			66-b 67-b
Okamura Mfg. Co., Ltd.		96-c 102-c 136-e	
Seiko-Seiki Co., Ltd.	112-d		98-c 112-d
Seilor Pen Co., Ltd.			125-e 126-e
Shinko Electric Co., Ltd.			65-b 105-c 106-c 133-e 134-e
Showa Kuatsuki Co., Ltd.			63-b 64-b 116-d 135-e 145-g
Star Seiki Co., Ltd.			107-c 108-c

Manufacturer	Electric Power	Hydraulic Pressure	Air Pressure
Sumitomo Heavy Industries, Ltd.		139-f	
Tokyo Institute of Technology	113-d	113-d	113-d
Tokyo Keiki Co., Ltd.		73-b 74-b	73-b 74-b 100-c 101-c
Tokyo University	42-a 44-a 45-a 77-c		
Tokyo University of Electricity	42-a 43-a 44-a 45-a	78-c	58-b
Tokushima University			55-a (Also: gas, 55-a)
Toshiba Co.			109-c 143-g
Toshiba Seiki Co., Ltd.			110-c 111-c 122-e 123-e 124-e
Waseda University	39-a 40-a 41-a 47-a 48-a	50-a 51-a 52-a 53-a	
Yamatake Honeywell Co., Ltd.	75-b		75-b
Yasukawa Electric Mfg. Co., Ltd.	57-b 81-c 82-c 83-c 142-g		128-e